The World We Know

Contemporary Science and Biblical Theology

PAUL ASHLEY

SeeLight
Publishing
Toney, Alabama

The World We Know
Contemporary Science and Biblical Theology

Scripture quotations taken from the
New American Standard Bible® (NASB),
Copyright © 1960, 1962, 1963, 1968, 1971, 1972, 1973, 1975,
1977, 1995
by The Lockman Foundation
Used by permission. www.Lockman.org

ISBN 978-0-9894452-1-4

Published by:
SeeLight Publishing
Toney, Alabama

Acknowledgements

The thoughts printed here have been circumspectly assimilated over the past 40 years. I wish to thank my wife, Joan, who through this entire time has been my continual support.

The illustrations were handsomely drawn or prepared by my sons, Brian and Stephen Ashley.

I would also like to thank Stephanie Anderson and Jason Orr of Jera Publishing for their insightful and patient assistance in the book preparation.

For with Thee is the fountain of life;
in Thy light we see light.

— PSALMS 36:9

Contents

Preface

Thousands of books, most of them larger than this one, have been penned over the past 50 years on the subject of science and theology or more specifically, science and the Bible. Among the various volumes is a wide spectrum of opinions. Theories, evidences, data, philosophies, as well as many controversies abound. No attempt is made here to enumerate a lengthy list of these or of their ensuing arguments, nor does there appear to be any point in doing so. They are readily available elsewhere. Some have questioned whether it is even useful or meaningful to discuss the subjects of science and theology conjointly at all, suggesting that they are mutually exclusive. Many others weld them together without regard to their curious and distinctive admixture. The intent of this small volume is to reflect insightfully upon the beliefs, purpose, motivation, and perspective which provides context for this ever interesting subject, regardless of inevitable new discoveries or advances in science.

Specifically, the theological confines of the work presented here are based upon the Bible and related biblical doctrine. The material is written with the theology or Bible interested person in mind. However, whether a student of theology or of science, the perspective which is offered should help each to understand the other.

Paul Ashley, DSc

The World
We Know

Introduction

I want to know how God created this world...
I want to know His thoughts,
the rest are details.[1]

— ALBERT EINSTEIN

T he scientist studies the world; the theologian studies God. There has forever been an interest in understanding the world as well as God among philosophers, theologians, and scientists alike. But it would appear that within the past two centuries, and even more so in recent years, theologians have become more interested in this question of understanding the origins of the world than in all of history. And the reason... because scientists have become so interested in figuring it out... *without God*, or at least without the God of the Bible. It is not so much that biblical theology is not compatible with science, but rather that contemporary science is often not compatible with biblical theology. Contemporary science along with the ensuing advances in technology would prefer to rephrase the question, "I want to know how *Nature* created this world". And so these two versions of the question frame the context of the subject which will be explored here ... without too many "details".

The discoveries of contemporary science are not so much what we did not know as they are what we did not know that we did not know. To illustrate, consider a scientist going for a hike along a trail in a nearby, well travelled nature preserve. He has been there many times before on the rocky paths which bear the impressions from thousand's of previous hikers. All is familiar to him. He knows the wildlife, the change in seasons, and even the sounds in the distance. But on this outing he takes notice of a large outcrop of rock not far from the path as he walks along. Examining it a bit more carefully his eyes are drawn to a dark patch on its surface. As he leans closer, he realizes that the dark area is actually a hole in an otherwise solid undisturbed stone. Curiously, he peers into the hole and sees that there is a cavity concealed behind it. With his pocket flashlight he is able to illuminate the interior through the hole and examine the walls. With his arm he reaches inside to investigate and feels the loose stones resting there on the bottom.

Before he has a chance to dwell too long on thoughts of his new discovery, his eye focuses on the far back wall inside the cavity. Straining to see, he finds that there is another hole in the distance leading to an even larger chamber filled with darkness. From his vantage point, with the flashlight focused carefully through the outer opening and onto the hole in the back of the cavity, he can only manage a glimpse of the space beyond. The column of light projects through the far away opening, illuminating only a small round area surrounded by a dark shadow, which all but conceals everything in the distance. And stretch as he might with his shoulder now firmly against the rock face and his arm aimed determinedly at the back wall of the cavity, this inner orifice is just out of the reach of his strained and extended fingers. What the casual scientist first thought was an interesting and pleasing discovery of a hidden chamber in a nearby rock, had now left him with a perplexing mystery far beyond his sight or grasp.

Likewise, decades of scientific exploration along the well travelled paths of the natural world have led to a growing list of mysterious chambers. For each there is often only a glimpse of what lies far beyond sight or grasp. And with each new mystery is the realization that there is far more that is not known ... and we did not even know it. You might say that compared to the fraction of our knowledge of the world that we thought we knew, then as a result of contemporary science, we actually know less. That is, we realize through the progress of science that the fraction of knowledge we know is getting smaller as we become aware of how little we know. And further, more and more of these new mysterious chambers of the unknown are each harder to attain than the previous mysteries of the past. For contemporary science at least, there are many mysteries which are perhaps beyond the sight and grasp of human beings. They are often left to speculative "scientific" investigation. For the biblical theologian there are mysteries as well, both of the natural world and the supernatural.

The term, biblical theology, is often defined as the doctrinal study of the Bible within its historical background, what was believed and taught through various periods of Bible history. In 1787 the German biblical scholar, J.P. Gabler, popularized the distinction between biblical theology and "systematic" theology, which considers the whole Text collectively to reach a particular doctrinal conclusion. Together the two concepts are somewhat complementary and aspects of both are intended within the broader use of the term here.

Since the world of biblical theology includes both the natural and the supernatural, the mysterious chambers beyond sight or grasp span a much larger context. Just as in contemporary science, we may find that in biblical theology the more we know invariably leads to the realization of how much more we do not know. Strangely, this is not necessarily an unfortunate circumstance. For the theologian, the more that is unknown or even unknowable, the more preeminent is the God of the Bible. The biblical Text puts it this way,

> The secret things belong to the Lord our God,
> but the things revealed belong to us ...
> *(Deuteronomy 29:29)*

Further, for the biblical theologian, the larger world context adds insights and perspective to each of the mysteries. Even so, speculative interpretation often remains just as much a temptation for the theologian as speculative investigation does for the scientist.

The author of a recent article in *Physics Today* enti-
tled "Thinking Differently about Science and Religion"[2]
expressed his unresolved quandary over what he called the
"conflict" between science and religion. He ended by citing
lines extracted from the Bible found in the Book of Job
along with his hopeful challenge for science to eventually
tame these intimidating questions,

> Have you understood the expanse
>   of the earth? ...
> Where is the way that the light
>   is divided ... ?
> Can you bind the chains of the Pleiades ... ?
> *(Job 38:18,24,31)*

The intent, however, of this pronouncement, made by
God to His servant Job, had a much different purpose. It
was to emphasize the vast incomparable difference between
God's knowledge and abilities in contrast to that of man.
Indeed, God's reply in these verses to Job began,

> Where were you when I laid the foundation of
> the earth? ...
> *(Job 38:4)*

Far from a challenge, it set Job back on his heels in awe.
He was speechless. And herein is the beginning of truly
understanding both the controversies as well as the won-
ders of biblical theology and contemporary science in *the
world we know.*

1

Nature's Mysteries

Gravity explains the motions of the planets, but it cannot explain who set the planets in motion. God governs all things and knows all that is or can be done.[3]

— ISAAC NEWTON

In 1968 the first Apollo mission to the moon was launched. Following a complete orbit around the earth and timed to coincide with a specific location of the moon in its orbit, the spacecraft's thrusters were activated to escape the grip of earth's gravity and eventually be captured by the moon's gravitational tug. The timing was designed to allow precise placement of the craft into an orbit around the moon with the addition of a few very brief mid-course firings of its small rockets later in the journey. This would be similar to hitting a target moving at a speed of 2288 miles/hour nearly 240,000 miles away from a platform that is moving in a circle at a speed of 1000 miles/hour and doing so while expending the very minimum amount of energy in the process.

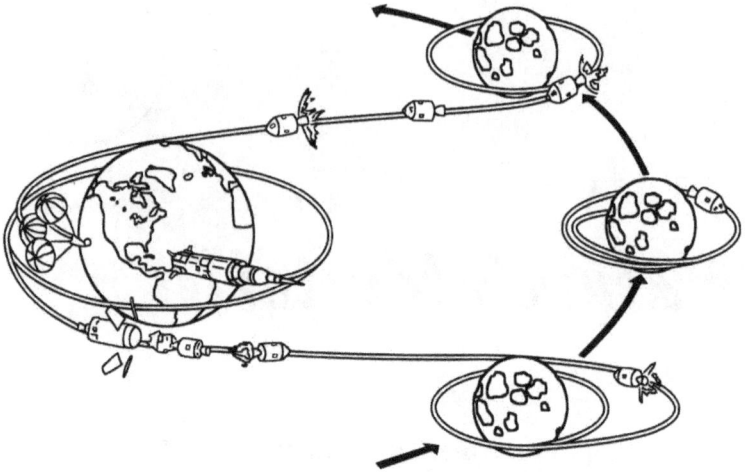

Apollo Lunar Mission

The spacecraft intersected a very narrow region just to the side of the moon, not so close as to be pulled in and crash on the surface and not so far that it may be ricocheted off, continuing forever into deep space. From there it orbited the moon 10 times before firing its thrusters, which propelled its escape from the moon's gravity, and then returned again within the earth's gravity resulting in a precise orbit around the earth before slowing for a descent and landing. An extra measure of safety was planned as the flight path was designed such that the gravity of the moon would slingshot the spacecraft around the moon in a *free-return trajectory* returning to earth wholly by the gravitation forces of the earth and the moon if an unforeseen rocket failure occurred. This contingency was put to the test when an oxygen tank suddenly exploded early into the Apollo 13 mission in 1970 forcing them to abort the mission and slingshot back to earth from the far side of the moon. When it was all over many would breathe a sigh of relief and thank God for gravity.

The Mystery of Gravity

Since ancient times people have recognized and used to their advantage the ever present force of gravity in the world around them. In the centuries following the Middle Ages its effects were methodically explored along with the aid of geometry. Eventually these ideas and their application within the natural world were modeled with mathematical precision, famously characterized by the brilliant philosopher and scientist, Sir Isaac Newton, nearly 400 years ago in his work, *Philosophiæ Naturalis Principia Mathematica*, (Mathematical Principles of Natural Philosophy or *The Principia*). Here Newton described gravity as, "... that force, whatever it is, by which the planets are perpetually drawn aside from the rectilinear motions, which otherwise they would pursue..."[4]

During the past century the concept of gravity has been an essential ingredient to nearly every development of modern physics as well as virtually all technological advancement, whether in the gadgetry of human life on this planet or in the exploration of space. And yet we do not know any more about the origin of gravity or why it exists than did the earliest civilizations of people on the earth. Even today the numerical value of the gravitational constant which determines the force of gravity between two masses must be determined indirectly and is only known to an accuracy of about 4 digits.[5]

Gravity's existence is merely presumed. It is presumed because it is wholly predictable, not only in the present but in the historical past. It is presumed rather than pondered because it does not beg an explanation. From the perspective of naturalism, upon which science is based, it has always existed and required nothing in order to exist. It requires no precursor. It requires no change or evolution to be palatable. You might say that it is accepted by faith both to exist and continue to exist in the future. As such very little if any thought has actually been given to the ultimate origin of gravity by most scientists. It is as if to recognize that such understanding is beyond reasonable expectation or necessity within the context of naturalism. Newton was content to say,

> And to us it is enough, that gravity does really exist, and act according to the laws which we have explained, and abundantly serves to account for all the motions of the celestial bodies, and of our sea.[6]

Even today science remains resigned to this same conclusion that gravity is presumed to exist without further explanation.

The Mystery of the Atom

By the same measure, three other fundamental forces of nature are likewise presumed to exist, electromagnetic, strong nuclear, and weak nuclear forces as they are commonly called. Electromagnetic force is responsible for giving firmness to objects in opposition to gravity as well as the electrical interaction which makes possible all modern electronic technology. For example when you sit down, it is the electromagnetic force of the atoms in the chair seat that holds you up, acting against the force of gravity. And it is an electromagnetic force which is harnessed to generate mechanical motion from an electric motor or to enable your electronic computer to function.

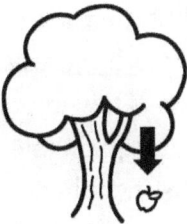

Gravitational   Electromagnetic

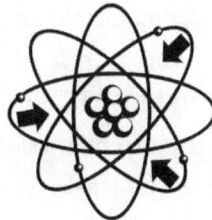

Strong Nuclear   Weak Nuclear

Fundamental Forces of Nature

The strong nuclear force keeps the nucleus of atoms from flying apart as a result of the internal electromagnetic forces of its constituent parts, while the weak nuclear force is involved with such things as radioactive decay. Each of these presumptions is interwoven along with many other presumptions to form the "rules of order" as it were for understanding the material world and natural phenomena. With these presumptions, models and theories have been formulated based on empirical observations. Out of this, such illuminating theories as general relativity and quantum theory were developed and later empirically observed beginning in the early part of the last century.

However, there is no scientific explanation for these four forces of nature, each very different. Gravity is much weaker than the other three but acts at a much greater distance and only attracts. Gravity acts on a mass but there are two kinds of mass. *Inertia* mass is how much something resists when you try to move it. *Gravitational* mass is how much something wants to be moved. But it is an unexplainable mystery of nature why these two apparently distinct phenomena are always *exactly* the same for a given mass. And so they are presumed to be identical, which is an essential assumption at the heart of Einstein's theory of general relativity.

Presumptions like these are generally considered comfortable enough to just be accepted in that they may be construed or interpreted in a way that does not challenge the notions of naturalism or determinism. But with the advent of *quantum theory* it became much more difficult to maintain this comfort zone. With it was brought to the forefront some surprising conclusions. The term "quantum" comes from the Latin word meaning "lump" or "packet" which refers to the concept that energy in its smallest quantity occurs in discrete values rather than as a continuum. One of these conclusions stipulated that at the smallest scale or dimension one cannot know both the precise speed and position of matter. These properties can only be known to within a certain probability, but not exactly known. This is called the Heisenberg uncertainty principle, named after German theoretical physicist, Werner Heisenberg, one of the pioneers of quantum theory. It would turn out to be one of several consequences of modern physics which were historically very disturbing to the rational naturalistic mind of the scientist.

The debate arose of whether this uncomfortable situation was the result of an elemental characteristic of nature that is actually unknowable or rather, whether it is merely not known or understood *yet*. Albert Einstein famously argued the latter while future proponents of quantum theory could and would accept the former. As Heisenberg proclaimed, "It will never be possible by pure reason to arrive at some absolute truth."[7] Much later physicist John Bell muddied the waters further with a thought experiment, known as "Bell's Theorem" which purported to prove the long standing proposition that a conscious act of measurement is necessary to "determine" the empirical outcome of a quantum defined system. That is to say that only when an actual measurement is made is the physical outcome no longer just an unknown probability of occurrence but rather a decisive result. And so the consequences of modern science had begun to collide with the basic scientific understanding of the natural world itself.

Presumptions or their conclusions were no longer conve-
niently benign. After all, how can a completely deterministic natural world not be completely determinable? An
undercurrent of unsettledness and debate has continued
as if watching someone shifting their position while sitting
uncomfortably on a distinct lump under the seat beneath
them, but at the same time trying to ignore it. Although
the formalism of quantum theory is generally accepted,
there are multiple interpretations of the implied enigmas.
One prominent speculation even suggests that there are
other parallel worlds in which the alternative outcomes are
played out simultaneously to reconcile all of the various
possibilities.

Over the last half century a large group of scientists worldwide have explored the realm of subatomic particles. In
1803 English scientist John Dalton proposed the concept
of the atom as a small round particle of which there were
no smaller parts. Over a hundred years later this picture
of the atom would be replaced by a model first formulated
by physicist Ernest Rutherford which consisted of a dense
small nucleus surrounded by electrons. But it was not until
1932 that another English physicist, Sir James Chadwick,
discovered that the nucleus was made up of positively
charged protons and uncharged neutrons, which are encir-
cled by electron orbits or clouds.

Although an understanding of radioactivity began in the
late 19th century, it was near the middle of the next century that the discovery of more elementary particles began

to proliferate. This was advanced with the development of ever larger and more powerful high energy particle colliders such as the one on the French-Swiss border (Large Hadron Collider, LHC) which measures about 17 miles in circumference. With the collider, tiny "bullet" particles (protons) are propelled with mass energy densities over 100 million times that of the atomic bomb dropped on Hiroshima. They are directed onto a target material resulting in the breaking apart of larger particles into sub-particles. But of the 12 more elementary particles of matter discovered along with 5 other non-matter particles that have been discovered, only 3 of them are associated with the formation of the components of matter, the proton, neutron and electron. The purpose of the others is unknown to science even as every few years new particles are discovered. And there is no scientific answer as to how many more particles remain undiscovered or what the ultimate most basic particles may be. There is no apparent end to ever more basic particles.

Cosmic rays are high energy particles which permeate all of outer space and bombard the earth continually. Some of them have energies that are over 10 million times higher still than the capability of any current particle collider. The energy of these cosmic ray particles are too high to explain by anything in the universe and their origin is completely unknown. It is believed that for these particles to have travelled far through the universe from some distance source, they should have lost much of their energy.

In 1928, English physicist Paul Dirac combined the young theories of quantum mechanics with Einstein's special relativity. This led a few years later in 1932 to the discovery of what would later be called *antimatter.* This is matter that is exactly the opposite in polarity to the matter in which the world we know and live in is constructed. The first discovery of antimatter was the positron which is basically identical to an electron but with the opposite charge. Matter and antimatter annihilate each other immediately when they collide. So antimatter is only rarely detected and disappears almost as quickly as it is "seen". There is no known scientific reason for its existence.

The list of mysteries of the natural universe for which contemporary science has no clue is very long. And with each new theory of the universe or its origin many more mysteries emerge from the theory itself or its inconsistencies. Here are a few more of those many mysteries.

How big is the universe? How much is beyond what we can see in outer space, which is limited by the speed of light travelling to us?

What is *"space"*? Do the laws of nature operate outside of space? Are they even the same within the distant space that we can observe since we will never be able to travel far enough to reach them?

What is *"time"*? Why does it have a preferred direction, only moving forward, never backward? There is no universally agreed upon definition for time which remains its own central mystery. Very few physics books even try to define it. But all the laws of nature are based on *causality* which means that present events are the result of past events and would not make sense otherwise.

Why is motion limited to the speed of light? This speed limit is just fast enough to see the stars but not fast enough to reach them. Is the speed of light the same in all of the universe and throughout all of time? The range of causality is set by the speed of light. The range in which things can happen within the speed of light is called our *locality*. Why does the concept of locality and causality appear to be designed for our particular human minds and the way that we think?

Why did the universe appear to start in such a highly organized configuration? The level of disorder is referred to by a term called *entropy* and always goes from order to disorder. On both large and small scales, whether among the bodies of outer space or the bodies of microscopic organisms, the progression of time is strangely linked to the inevitable increase in entropy.

Why is the universe set to a perfect "age" and "balance" at this time of human existence as some scientists call it. Gravity holds massive shapes together such as solar systems, galaxies or super-galaxies. And yet they are not pulled apart by the other energy within the universe.

In a contemporary book exploring the current status of our scientific understanding of the universe and its origin through the study of cosmology, physicist Daniel Whiteson repeats ... "We have no idea" ... or some similar expression over 70 times. Although convinced of an ultimate naturalistic explanation, he concedes, "just how absurd it is to think that we have any clue what's going on or how the universe really works."[8] He writes, "we are still in the dark about most of the basic truths of the universe."[9] And he goes on to say regarding the countenance of contemporary science,

> More recently, this feeling has been replaced by a cool, casual confidence ... — the feeling that the world around us can be described by rational discoverable laws.[10]

Then he adds,

> It seems our mastery applies to only a tiny corner of the universe, and we are surrounded by a vast ocean of ignorance.[11]

The Mystery of Life

There are of course many other things about the natural world that we do not know. And for the same reasons they are difficult to encompass within the realm of a physical world governed by naturalism. One of them is the distinction between living and non-living material which has always been a mystery. *Vitalism* was the term given to the concept that there is an inexplicable non-physical element which governs living organisms. By the 1930's, with the strong impact of naturalism and the development of organic chemistry, the concept of vitalism was no longer widely recognized by science.

A mechanistic approach to the development of a living organism was assumed based on the understanding of cells and genetics. The concept of *reductionism* was further applied in which at the lowest level life could be defined strictly by natural chemical processes. But nearly a century later the mystery of life in living material still remains far beyond the grasp of science. We can preserve, modify and resuscitate living material. We can even examine living material down to the very atoms of which it is composed, but we cannot fabricate or generate a living organism from non-living material by naturalistic science.

The definition within biological science of a living organism compared to a non-living organism is rather artificial and somewhat arbitrary. Some of the attributes often designated for a living organism are respiration, intake of energy, reproduction, and sensitivity to environment. But the term "living" is conspicuously buried within the larger concept of an "organism". The description of "living" is a state or condition, not an organism. It is simply the opposite of "dead" and a very fine line separates the two. In a multi-celled organism the concept and understanding of "living" becomes even more elusive. One could argue that in general all of the parts of a living organism are capable of being alive, but at some point parts may be alive and other parts no longer alive.

Somewhere within the dynamics of cellular chemistry reductionism begins to fail and a naturalistic description for this mystery is left wanting. At some point cellular chemistry faces a "singularity" of sorts which for the naturalist is, at least for the time being, an acceptable oversight. In 1944 Austrian physicist Erwin Schrödinger, who developed the fundamental equation for quantum theory, struggled with the dilemma. He wrote in *What is Life?*

> From all we have learnt about the structure of living matter we must be prepared to find it working in a manner that cannot be reduced to the ordinary laws of physics.[12]

Now over 60 years later, having dug deeper with molecular and genetic level discoveries, for contemporary science the mystery of life remains elusive.

Philosopher C. H. Lewes coined the term "emergent" to describe what Aristotle had referred to as the whole being greater than the sum of its parts. He wrote in 1875, "The emergent is unlike its components insofar as these are incommensurable, and it cannot be reduced to their sum or their difference."[13] Contemporary science has over the years adopted a weaker form of the term, in which the whole is always somehow naturalistically reducible to its components, albeit completely unknown. So the label, *emergent phenomena*, is now often applied equally to a marvel such as a snow flake as it is to the mystery of life itself.

If life is a mystery, than far more of a mystery is the distinctness of human life and consciousness with its many unique attributes. The human brain contains several billion neurons. They appear to perform a simple function of relaying electrical impulses across synapses to each other. Out of this physical phenomenon there arises the conscious mind. It cannot be reduced to the mere interaction of neurons.

In contrast to all other forms of life, humans are characterized by the ability to reason, to comprehend time or events in both the past as well as the future, and to exercise free will. The understanding of these traits is not at all known from a naturalistic perspective. However, they may be empirically observed in a similar way as gravity or some other phenomenon is observed. Together, these qualities of human life challenge the law of determinism which is fundamental to the natural world.

The notion of free will requires that the laws of nature at some level become compliant to a mental choice. More directly, the laws of Nature, *as we know them*, must be superseded. Nature must allow for that which is not predictable, or choice is not really a choice. And if nature allows for that which is not natural, then every act of choice is supernatural. Otherwise it must be presumed that there is no actual free will and every thought or action of choice is really totally deterministic and its outcome ultimately predictable. Obviously, this is not a very palatable alternative, leaving a considerable unresolved dilemma for naturalism.

Many of the mysteries of nature go almost unnoticed by both scientists and theologians as well as by most others. We are generally complacent about such things, having lived long without the need to resolve such conundrums. But it is these puzzles of the cosmos which may draw our attention to the difference between the world of the scientist and that of the theologian *in the world we know.*

14

2

Different Worlds

We comprehend the earth only when we have known heaven. Without the spiritual world the material world is a disheartening enigma.[15]

— FRENCH WRITER JOSEPH JOUBERT

In 1933 while observing a cluster of galaxies called Coma, Swiss astronomer, Fritz Zwicky, stumbled across a puzzling discovery. The Coma Cluster consists of approximately 1000 galaxies spread over an area roughly the size of a thumb held at arm's length. While perched high on Mt. Wilson Observatory, Zwicky used a special telescope designed to capture the entire cluster in a single photograph. There he collected data which he used to calculate the movements of the galaxies based on their light spectra and then estimated the gravitational mass of these galaxies based on their complicated orbits. He discovered that the total mass was 400 times greater than all of the mass that could be accounted for by the radiated light from the stellar bodies comprising these galaxies. From this he inferred the existence of *dunkle Materie* or "dark matter".[16] This dark matter along with dark energy represents what is popularly believed to be unexplainable matter and energy necessary to reconcile the physical motion and distribution of the heavenly bodies across the cosmos. Without it the observed magnitudes of gravitational attraction and interactions of these bodies would not make sense based on current theoretical understanding or assumptions. There is something apparently missing. It is suggested that this missing material is characterized by having no electromagnetic interaction with normal matter and energy. Therefore, it does not give off light, hence the name, dark matter. Today it is believed that dark matter and dark energy make up about 95% of all matter and energy in the universe. And dark matter represents about 85% of all matter. In other words, the vast majority of matter in the universe is not visible or detectible by any normal means. One could imagine that if it were suddenly illuminated, a vast universe, far greater than that which we know, would appear before our eyes. It would envelope our bodies and our surroundings. It would fill the sky and stretch into distant space in all directions.

It would be over 50 years later before contemporary science was able to explore this phenomenon more fully. In recent years scientists have attempted to detect dark matter, hoping to observe a rare collision with a small particle of normal matter. So far there have been no conclusive results. What remains a startling puzzle for contemporary science is strikingly similar to that of the spiritual world. Consider for comparison a description of dark matter: a vast cosmos, both near and far, within reach and yet out of reach, knowable and yet invisible, many times greater than our own world which we know so well.

Collision of Presumptions

The collision of presumptions associated within natural-ism is not new. The theological picture of the world and that of natural science have for a long time been painted from a completely different palette. And the two have collided many times. Within the past century and a half this difference has become considerably more pronounced. The world of natural science is strictly bounded by what can be observed with the five senses. It is confined in both time and space to that which is connected and determined by laws within human reason and comprehension. It is a physical world defined by that which the human body has a cognizant, natural ability to physically sense and mentally comprehend. Well known British science writer Nigel Calder expressed it this way,

> What men make of the universe at large is a product of what they can see of it and of their own human nature.[17]

The theological world, described by the Bible, on the other hand, is not limited by such bounds. It acknowledges a reality beyond that which the natural human senses can detect. And it recognizes events, actions, and consequences which are not consistent with natural laws or derived from empirical observation. In the Greek language there are two words which describe and distinguish realms of the world. One is *psychikos*, which is often translated "natural". It means of the five senses or sensual. This relates to the world that one knows by virtue of our physical connection to it. The other word is *pneumatikos*, often translated spiritual, which means relating to the supernatural. This is associated with the world that is beyond the bounds of our physical senses. It is realized by way of altogether different "senses" that extend beyond the natural. From the biblical Text,

> But a natural man does not accept the things of the Spirit of God, for they are foolishness to him; and he cannot understand them, because they are spiritually appraised.
>
> *(1 Corinthians 2:14)*

Here the word translated "appraised" is *anakrinetai*, which means to examine, often used in a forensic sense, to methodically investigate and understand. It is present tense, passive voice which indicates a continual ongoing action of the spiritual "senses" upon the object of investigation. This is analogous to the role of the physical senses.

The world of biblical theology is inconsistent with that of naturalism in several specific ways. The biblical description of human beings endows them with a uniqueness when compared to all other living creatures in that they have a free will, capability to reason, and a soul which is immortal, living on beyond the death of the physical body. This life after death implies a spiritual realm, beyond the bounds of the natural senses and the natural world. Within this spiritual realm is the abode of other creatures and parts of God's larger creation such as angelic beings or heavenly dwelling places. The whole of both the natural and spiritual world is under the divine influence and control of God. And God's intervention in both realms supersedes all laws of nature and other influences. All of these elements of biblical theology run counter to the world of naturalism. As Henry Thoreau once opined, "With all your science can you tell how it is, and whence it is, that light comes into the soul?"[18]

Therefore, it should be no surprise that many of the concepts of modern science such as quantum theory are also difficult to reconcile under the constraints of naturalism. And it is equally no surprise that natural science often collides with the theological world for similar reasons. There is no reason to believe that as modern science continues to explore the physical world more deeply that it will not repeatedly stumble upon these uncomfortable "lumpy" boundaries which defy the criteria for naturalism and determinism, and thereby challenge naturalistic human reason.

Perceived Differences

Aside from the obvious differences between the natural and spiritual world, there may be other perceived differences which often result in confusion. Some of these stem from a historical context. The understanding of the natural world in which we live has been broadened over the centuries as a result of exploration and accumulated knowledge. At times religious institutions have been closely associated with the cultural and political management of philosophy and the development of science which emerged from this discipline. Under such control current philosophical and scientific knowledge of the time was vested with religious authority and interpretation. This led to continuing controversies as the understanding of the present natural world changed. For example, the geocentric model of the world with the earth at the center was challenged in the 16ᵗʰ century by Copernicus and others. The heliocentric model with the sun at the center undermined the religious institution's concept of their earthly domain being the center of the universe. Another concept of the natural world with historical significance was the understanding of the shape of the earth which in many ancient civilizations was believed to be flat and surrounded by water. But it is worth noting that the history of the flat earth is often misrepresented. It is widely believed that the concept extended into the middle ages. Actually, the belief of a flat earth began to fade at least by the 6ᵗʰ century BC as evidenced by the Greek philosopher Pythagoras[19], and the concept of a spherical earth was already commonplace by the time of Aristotle in the 4ᵗʰ century BC.

But these historical developments in the understanding of the present natural world are not part and parcel of biblical theology. The Bible does not advance the notion of a flat earth or a geocentric universe any more so than modern day language which refers to the 4 corners of the earth or the sun rising in the east. To the contrary, modern concepts of the natural world are found throughout the Text in spite of having been written centuries ago. There is no Hebrew word for sphere but the word *chug* which is often translated "circle" comes the closest to a word for describing a sphere as found in the biblical Text of Isaiah 40:22, "It is He who sits above the circle of the earth…" Biblical theology, distinct from religious and secular embellishment, is consistently found to be ahead of, not behind in the progressive understanding of our *present* natural world.

What is generally called contemporary or modern science emerged during the early part of the 20th century. From this time the term classical science came to be associated with all prior science. Among the very significant foundational developments of classical science during the 18th and 19th century, which remain in use today, are the principles of mechanics along with thermodynamics and electromagnetism. In contrast, the epic principles of contemporary science were discovered over a relatively brief period of a few decades, culminated by the 1930's. Contemporary science since that time has simply built upon those principles and still today references them foundationally as various theories are extended. Similarly, modern technology in all of its vast diversity is enabled by these same principles of the early 20th century.

Unresolvable Beliefs

Intelligent design is the term often invoked to describe the observation that the elements of nature as they appear in the world are not possible without prior direction and control. The controversy over the origin of the world finds its roots here. One may choose to believe either in intelligent design or random evolution, or even random evolution enhanced by natural selection. But neither intelligent design nor evolution can be proven by traditional scientific investigation. This would require observations that are repeatable and verifiable by experimentation. So there will forever be a controversy here because proof by the traditional scientific process is not possible. A decision emerging from one's belief is required.

The crux of the issue is well-set with the illustration conjectured by the English clergyman and Christian apologist, William Paley in *Natural Theology*, 1802,

> [S]uppose I found a watch upon the ground, and it should be inquired how the watch happened to be in that place, I should hardly think ... that, for anything I knew, the watch might have always been there. Yet why should not this answer serve for the watch as well as for [a] stone [that happened to be lying on the ground]?...[20]

The Watchmaker

Since the time of its publication Paley's argument has been quoted often and has prompted much discussion or debate. It distinguished between the two objects in that unlike the stone the watch is perceived, with its various pieces, to be framed and put together for a purpose. This *teleological* observation was evident in the Bible centuries ago,

The heavens are telling of the glory of God;
And their expanse is declaring the work of
His hands.
(Psalms 19:1)

Spanning time it is found again in the admonition
addressed to those who were most certainly influenced by
the Greco-Roman culture long ago in the first century, lest
they would doubt,

For since the creation of the world His invisible attributes, His eternal power and divine
nature, have been clearly seen, being understood
through what has been made...
(Romans 1:20)

The words of Werhner Von Braun, one of the leading
figures in rocket development for space exploration, whose
experiences with science led him to God, demonstrate that
despite the objections of some this argument remains a
convincing one,

They challenge science to prove the existence
of God. But must we really light a candle to
see the sun?
What strange rationale makes some physi-
cists accept the inconceivable electrons as real
while refusing to accept the reality of a Designer
on the ground that they cannot conceive Him?[21]

Naturalism rejects the notion of intelligent design but replaces it with an equally faith-based concept. The presence of design is inferred only to the extent that an object's characteristics differ from nature. To the naturalist nature is the reference for that distinction. As one prominent writer in support of a Godless nature explained,

> Therefore, to claim that nature as a whole was designed is to destroy the basis by which we differentiate between artifacts and natural objects.[22]

This is very much like the teleological argument but based on the assumption or *belief* that nature is simply _not_ by design. This argument is fundamental to naturalistic science when applied to origins of the world. Popular evolutionary biologist, Richard Dawkins, writes in defense of evolution, "Even if the evidence did not favour it, it would *still* be the best theory available [italics mine]."[23] Actually, it is the only theory available for this chosen *belief.* The well known science fiction writer and ardent evolutionist, Isaac Asimov, commented, "I'm a creature of emotion as well as of reason. Emotionally, I am an atheist."[24]

Scientific study of present day nature and the world around us does not necessarily compel a choice in one belief or the other. Scientific investigations which do not extrapolate with inferences to the distant past are generally applicable to the traditional scientific process of verification. This applies to the vast majority of scientific study. For this reason it is possible to find scientists of all faiths and beliefs with little conflict in many of the disciplines of science and engineering. This is particularly true in the sciences that support present day technologies in which the purpose and application is confined to the present or limited future. But in those areas that border on cosmology and life sciences it would be hard to avoid a conscious decision, one way or the other, regarding intelligent design and the universe, as a matter of conscience. These disciplines tend to be more identified with current popular theories relating to world origin and incorporate these theories across most of their discipline areas. There is inevitably an increased intersection of the natural world with the spiritual world of the Bible as one proceeds further into the past. And therefore, it is not surprising that when it comes to the subject of nature's past there is much controversy in belief.

For human beings at least, all understanding of truth hinges on faith, is acquired by faith, and is accepted by faith. This applies whether it is theology or science. With the theology of the Bible it is stated clearly without disguise. With science this requirement is in the fine print … which many scientists do not read. For the naturalistic scientist or for the biblical theologian, their faith defines *the world we know.*

Boundary of Science

God does not suffer presumption in anyone but Himself
— GREEK HISTORIAN HERODOTUS

T he land of Egypt was considered an ancient civiliza-
tion in the time of Abraham. It abounded with large
scale engineering, technology and medical prac-
tices. Pyramids, temples and statues dotted the landscape.
From as early as 3000 BC this knowledge and scholarship
along with its many discoveries was recorded in a unique
hieroglyphic language. By the end of the Roman Empire,
hieroglyphics were no longer used and the understanding of
this language was almost completely lost even by Egyptian
scholars. As a result much of the outstanding Egyptian history, knowledge and advancement became a mystery, locked
away in the past. Scientific investigations even in modern
times could only infer the answers to many questions such
as the history of the pyramids, how they were constructed,
and the technology developed or incorporated in them.
Over the years a multitude of scientific theories abounded,

which have waxed and waned in popularity. As if to tease any visitor to the land, the countless hieroglyphs covered the walls and filled the aging scrolls of papyrus. The written record, bringing attention to its own immeasurable value, stood silent, awaiting the decipherment of the long forgotten language.

During the following centuries since about AD 400, many attempts to decipher the hieroglyphs or to understand the language failed and many proposed explanations of the alphabet led to misguided or false results in understanding. At the end of the 18th century one campaign of the French military and political champion, Napoleon, made its way to the northern coast of Egypt. In July of 1799 a couple of miles from the small town of Rosetta, which is just east of Alexandria, a French soldier, while digging a trench, unearthed a large stone fragment. It was inscribed with Egyptian markings from the early second century BC. Preserved there was a parallel inscription in three languages from the time of King Ptolemy V. Included was both the Egyptian hieroglyphic script and a shorthand version of the hieroglyphs, called Demotic script, as well as ancient Greek.

Working from this invaluable inscription, slowly by the 1820's an ancient Egyptian grammar and a hieroglyphic dictionary had been completed, identifying the phonetic elements and ultimately deciphering the language. With this, many of the mysteries of the Egyptian past including its history and technology could now be accurately understood from its own writings. One of the remarkable finds from the hieroglyphic language was their advanced mathematical development which was one of the earliest, if not the earliest use of a decimal system. No amount of scientific investigation alone could have resolved with complete confidence many of the mysteries of Egypt. Only a written record could with remarkable clarity open the window to Egypt's past.

These mysteries of Egypt's ancient history represent only an inkling in comparison to the prehistoric past of the universe. Aside from theories and inferences, the universe of the much more distant past is locked away. We only know *for sure* what happened yesterday or last year or the last century because of an actual witness or the existence of a reliable recorded document. And for the prehistoric past, the only witness would be God and the only reliable recorded document would be His. The untold details of Egypt's past that are not recorded in the ancient hieroglyphs may remain a mystery subject to speculation and wonder. And for biblical theology, God's untold details of the cosmos' past are likewise mysteries left to wonderment, but should wisely not be given to speculation.

Extension of Paradigms

It would be enough if we just stopped here and simply focused the study or understanding of science on that which is empirically observable and verifiable in the natural world, as did those such as Isaac Newton and other earlier philosophers of nature. A very common rationalization is to relegate science to the "how" and theology to the "why". For exploring the present natural world this may to some seem reasonable, but for the past, for issues of beginnings, understanding the "why" is indispensible to deciphering the "how". Students of modern science have chosen to extend conclusions beyond that which is empirically observable. And consonantly, students of the theological world have often countered these efforts with the application of similar naturalistic reasoning to explore beyond that which is historically observable. Among the greatest *presumptions* of all are the ones which exceed the bounds of reliable historical record and are not empirically observable or verifiable.

The risk here is that these necessary presumptions sometimes require further ones to avoid contradictions or collapsing altogether. For example, following the supposition of a self-evolved natural world, the question arises, what pre-existed the natural physical world? This is of course not known any more than is known, "What is gravity?" If fact, you might say that the knowledge of one may even depend upon the other. One popular presumption is that the natural world and the laws of naturalism have always existed. Therefore, any knowledge of preexistence is unnecessary. The natural world itself is simply a presumption which needs no precedent, prior knowledge, or ... *creator*. But naturalism also requires that the physical world remains unchanged even if these presumptions were not so. Further, the physical properties, rates and conditions must essentially have remained unchanged as well as natural laws, both spatial and temporal. Here we have the foundational underpinning of naturalistic reasoning. This concept is sometimes known as "uniformitarianism". The supposition along with its repercussions is not new as seen in the writings of the early apostles in the Bible,

... mockers will come ... saying, " ... all continues just as it was from the beginning of creation."
(2 Peter 3:3,4)

There are recognized today 19 independent fundamental constants of nature such as the gravitational constant or the charge of an electron, which is exactly the same but opposite charge as a proton. These constants are so finely tuned that if they were even slightly different in value, the universe would be so radically different that it would not support life. This is known as the *anthropic principle* and has led naturalists to consider a variety of interpretations to rationalize the existence of the universe without the need for a creator. Among these is the concept of a multiverse in which ours is one universe of many which happened to have just the right combination of physical parameters, however miniscule the probability.

Traditionally, the concept of science or scientific investigation from the time of Newton was confined to theoretical models or conclusions which may be validated either by empirical observation or historical data coupled with experimental verification. This process of investigation came to be called the *scientific method*. Generally, this was expected to involve both inductive and deductive techniques. In the twentieth century the concept of *falsification* was advanced by the famous philosopher of science, Karl Popper, to further clarify the connection between theory and experiment. In order for a theory to be tested it needed to be falsifiable. By this it is meant that a verifiable condition exists which could be observed or measured to invalidate the theory. Thereby the scientific method was considered applicable if there was a way to prove the theory to be false. Any investigation which is not supported by these elements and is not falsifiable or extrapolates in time or space beyond such validation would not be considered science. It would be considered speculative and sometimes the reasoning, methods, or conclusions which are involved may be recognized by another term, such as *pseudoscience*.[25]

Later in the twentieth century this definition of science was relaxed to include such disciplines as those defined by social scientists and behavioral biologists for which it was more difficult to establish falsification and experimental verification. The term "theory" become less distinguishable from the term "hypothesis", which in turn is less dependent upon verifiable observations. As an extension of this broader definition of science, a theory could, by popular consensus, become accepted and given the status of "paradigm" without the need for experimental verification. This term has notably become very much more frequent in use over the past few decades. Physicist Fritjof Capra described a paradigm as, "A constellation of concepts … shared by a community, which forms a particular vision of reality".[26] ***Simply put, a theory that is not falsifiable requires faith to believe and a paradigm is a popular theory adhered to by the faith of many.***

As paradigms became authoritative within scientific thought, the supporting theory and observations became justified as valid examples of the scientific method. It no longer mattered that theories involving the past or far expanse of the universe, beyond human reach, were not testable or falsifiable. The definition of the scientific method was extended, accommodating the fitting of observations in the present time and current locality to theories applied to the past and distant localities. And the term "testable" or "provable" was redefined to mean that the observations fit a theory regardless of the uniqueness of the theory or the assumptions required.

Over time this popular paradigm based philosophy became prevalent at least in part within the disciplines of science such as physics, chemistry and biology as well as the social and behavioral fields. It influenced scientists and philosophers alike. The clearest examples in the physical sciences are those which explore the prehistoric natural world in which experimental verification and therefore falsification is not possible. Three scientific areas of study within the scientific disciplines that were most strongly influenced by this paradigm philosophy were geology, zoology, and astronomy. Included in this list of popular paradigms were the theory of evolution and the theories of the origin of the universe.

Cosmology

Historical records go back at best less than 5000 years and most accurate physical data less than 400 years. Therefore, theories of the origin of the universe extrapolate into the past from these by a factor of over 2,000,000 times beyond verifiable parameters in time and its related historical setting. Presumptions are made not only about the constancy of many physical parameters over that vast period but also the validity of natural laws under extreme conditions of temperature and density which are not possible to replicate on earth or in a laboratory. For many of the physical parameters even a small change could result in extremely different conclusions due to sensitivities, instabilities, or inconsistencies in the theory. Some of these inconsistencies, for example, involve irreconcilable results from the expansion of the universe in the initial period following the "Big Bang", or the initial point of origin for all matter.

The prevailing scientific narrative for the origin of the universe — the Big Bang — goes something like this. At a particular point in time, currently believed to be 13.8 billion years ago, the creation of the world occurs. The prior state is not only unknown but non-existent. The event occurs spontaneously, without a creator or any pre-existent condition. For the smallest instant of time the entire universe is extremely dense and hot, infinitesimally small. But then it rapidly begins to expand and cool, all of this in about a billionth of a trillionth of a trillionth of a second. During this very short-lived period, understatedly called "inflation", both space and time (called space-time) expand enormously out of nothing, becoming very uniform everywhere. But at the same time tiny fluctuations (caused by quantum uncertainties), generated unpredictably in the process, remain. These fluctuations will eventually define all of the structure of the universe from the largest galaxy to the microscopic composition of the smallest pebble. Within seconds the

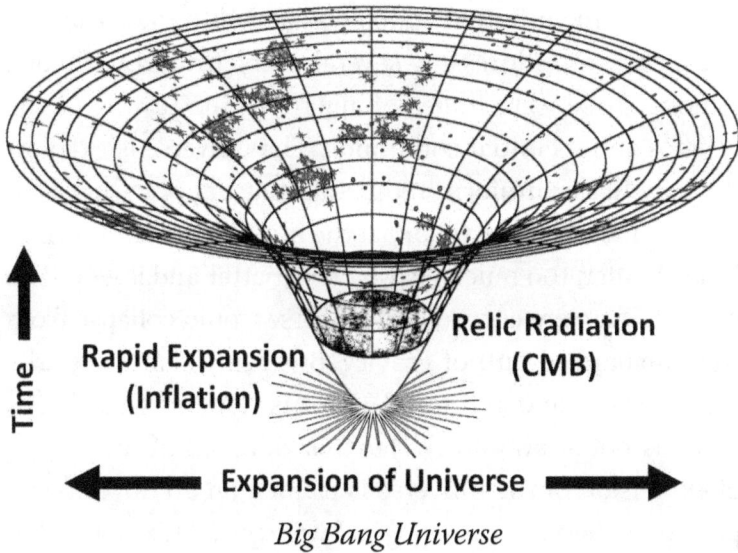

Big Bang Universe

temperature has fallen to a few tens of billions of degrees and the universe is composed of various random particles which will eventually contribute to the detailed formation of every object in the observable world. But in addition, the vast majority of these particles will form an invisible host of matter and energy that fills and comprises the entire universe, whose only evidence of existence is its apparent gravitational attraction among the stars.

After a few minutes have passed, the lighter elements such as helium and lithium begin to form. About 400,000 years later the Universe has cooled to a few thousand degrees, and light energy separates from matter in a process called recombination. This allows gravity to cause matter to collapse into the stars, galaxies and clusters of galaxies which are observable in the universe today.

The cosmic inflation epoch was added to the basic scenario early on, addressing several problems in the theory. Among these were issues of apparent fine tuning of the cosmos and its initial conditions. The universe is extraordinarily flat or uniform requiring unusual preciseness in the parameters of the theory at the inception of the cosmos. In particular, too much or too little matter and it would be curved. If it were curved, the universe would collapse from the imbalanced pull of gravity. But being flat is actually a very unstable condition. Why it is flat and what keeps it flat is not at all known. Furthermore, all of the theory of expansion of the universe is framed on an unverifiable premise called the *cosmological principle*. This states that the universe is homogeneous and isotropic when viewed on a large scale. In other words it presumes that we are not at a special place in the universe.

Among the difficulties with the theory is that patches of sky on opposite sides of the sky appear to have nearly the same temperature. But they would have been totally outside of any possible physical contact because they are separated by a time greater than the presumed age of the universe. However, an even more disturbing problem is that as the theory is extrapolated backward, close to the time of "creation" the physics model used which is based on general relativity is expected to break down. At some point a new quantum physics model would be needed because a mathematically impossible situation (called a singularity) would exist very close to this initial point in time.

Also there remained the dilemma of where all of the structure that we see in the universe comes from. So, the concept of a "vacuum state" was conceived in which the embryonic universe is composed of a strange fluid which exponentially expands over a very short epoch. Violation of quantum physics is then avoided. Small quantum fluctuations of the right form and kind are assumed to be generated from inflation which explains the creation and evolution of all of the observed structure in the universe as matter coalesces under gravity during the subsequent slower expansion of the cosmos.

While "fixing" some problems with inflation other concerns remain unsettled among cosmic scientists. To the point, consider the words of renowned cosmologist Robert Sanders, who happens to hold no regard for the notion of a God inspired creation of the universe. Although he maintains a steady hope and faith in science that a palatable naturalistic explanation will eventually emerge for the origin of the world, he recently wrote,

> It can be justifiably argued that inflation is not in fact a theory (there is no proper microphysical theory behind the idea) but a paradigm — a wish list of what we would like inflation to accomplish.[27]

And as such this concept is implicitly accepted by most cosmologists although what caused it is totally unknown.

So debates continue regarding the Big Bang theory which is not without controversy in spite of its widespread acceptance. Most recently, renewed concerns have arisen suggesting that the speed of light, the most important of physical constants may have been faster at times far in the past. Furthermore, our understanding of gravity completely breaks down at the moment of the Big Bang. The general theory of relativity depends fully on the constant speed of light as well as all the laws of nature including gravity, electricity, and sub-atomic forces. Indeed this, "One special feature of the cosmos has a grip on all the operations of the atom, star and quasar."[28] And these presumptions are key to all projections regarding the universe's past. Supporting data such as *cosmic microwave background radiation* and its interpretation is also dependent on these assumptions. With just a slight variation in the balance of the model for the expanding matter and energy, the universe would be rapidly collapsing or expanding.

Cosmic microwave background radiation (CMB) is the low level background radiation which permeates space in all directions. It is electromagnetic radiation at microwave frequencies and corresponds to the energy given off from a radiation source that is very cold, about 2.7 degrees above absolute zero. It was first detected serendipitously by radio astronomers while testing an antenna in 1964. Scientists were quick to make the popular association of this observed radiation with the expected residual radiation from the formation of atoms described above as electrons are combined with protons in recombination. However, numerically, the predictions for CMB over the years were very widespread and the closest was about a factor of two from the actual observed temperature. The energy given off is believed by cosmologists to have travelled through space to the present time, with the wavelength and corresponding temperature reduced over a 1000 times to a very low value. This radiation data collected by satellites over the past few years has been fitted to the theory and other numerical parameters were adjusted with the assumption that it is indeed from the Big Bang. As Sanders observed in his field of cosmological study,

> ... [the concept of dark energy and dark matter] has become something of an official religion, outside of which there is no salvation and beyond which there is only damnation ... a dogma that is too widely and too deeply accepted.[29]

The invention of "dark matter" and "dark energy" play an essential role in the theory related to CMB. Although there is no direct observation of them, they are inserted into the theory to explain the tiny variations in the background radiation. And further, dark matter and dark energy are employed to show how these variations, in the assumed proper ratio with real matter, can create all of the vast detailed structure of stars and planets seen in the universe. This includes the unique part of matter which evolved into life on earth and perhaps elsewhere. And thus it is, that the so called *standard model of Big Bang cosmology* just described connects these three.

CMB has been cheered as the Rosetta stone of cosmology. But the true Rosetta stone was actually a historical record, engraved by human hands. It contained prescribed intelligent information which provided the direct knowledge to decipher human historical documents and other writings. It was these human scripted records which unlocked mysteries from the ancient past with clarity and dismissed many speculations. On the other hand, CMB represents merely a present day observation of natural radiation. It is not a code to unequivocally translate past written historical records. At best it provides some empirical data which, if combined with many assumptions, may be inserted into various mathematical models. It cannot validate or prove a theory. There will always be other possibilities for interpretation. It is to be sure, no Rosetta stone.

Due to the numerous assumptions made and unverifiable ingredients in the CMB model there are several popular alternatives among cosmologists at the present time. Some of the most popular of them simply consider that the laws of Newtonian dynamics may not apply in the same way outside the local range of the cosmos. Therefore, they may need to be modified at the cosmic level of galaxies and larger, considering the puzzling observations of inconsistencies in the orbital speeds of stars in spiral galaxies and the elusiveness of dark matter. Considerable debate has ensued. However, it highlights the point that not only do we not know if the laws of physics were the same in the distant past, but we do not know if they are the same at the galactic or cosmic level. And there is no falsifiable way to test them.

The Big Bang theory suggests a point in time for the origin of the universe. But it does not address the question of what was before the Big Bang. It merely presumes that the universe is capable of self generation and leaves it at that. This is considered acceptable within the scope of naturalism, just another lumpy issue. It was not that many years ago when competing theories prior to the paradigm of the Big Bang suggested either an infinite time history of the universe or an oscillating universe which periodically expanded and then collapsed, only to be repeated over and over again.

In 1929 astronomer, Edwin Hubble, for whom the Hubble space telescope is named, was the first to observe and document the "redshift" of light from stars outside our own galaxy. The redshift refers to a very slight change in the wavelength or color of starlight. In this case the shift is toward the red end of the spectrum in the spectral signature of the gases of which the star is assumed to be made. Using the 100 inch Hooker Telescope at Mt Wilson, the world's largest at the time, he was able to show that the redshift was greater with greater distance to the stars. Other scientists were quick to notice that this result, when combined with a particular interpretation of Einstien's theory of Relativity, suggested that the universe was expanding.

The result was quickly extended to embrace the consequence that the expansion had a beginning. Anticipating such a conclusion, Hubble resisted the notion throughout his long career that the universe was not infinite and referred to his own prior observations as an "apparent redshift", believing that there may be some other possible explanation. The debate continued on and the term Big Bang was first coined by the leading astronomer Sir Fred Hoyle on BBC radio's *Third Programme* broadcast in 1949.[30]

In the program on cosmology Hoyle actually used the term as a critical remark while defending his alternative Steady State Theory. By the mid-1960's the Big Bang theory had achieved popular acceptance.

But the oscillating universe concept has now received some new interest. One of the recent proponents of this concept, which he calls the "Big Bounce", is cosmologist Neil Turok, Director of the Perimeter Institute for Theoretical Physics in South Africa. Speaking of the currently popular cosmological models he said, "In my view, these are all too complicated and arbitrary and contrived."[31] And so the theories of naturalism are ever changing as the paradigms shift.

Origin of Life

The presumption of random mutations alone leading to successively more complex outcomes as a basis for the theory of the origin and evolution of living organisms has been shown to be completely unsupportable by arguments of probability and statistics. Based on probability studies, evolution by random mutation has been shown to be mathematically far less than 1 out of 10^{50} (1 followed by 50 zeros), which is considered mathematically to be an impossibility. Regardless of the age of the earth, natural laws including those of thermodynamics, are inconsistent with the increased complexity as needed in evolution by random mutation, over any length of time.

More recently, recognizing the dilemma, some naturalistic theories began to suggest that instead of a purely random process, perhaps "natural selection" must somehow aid with the elementary stages in the process of "abiogenesis" as well as in subsequent evolution to increase the probability. Abiogenesis is the term used to describe the original incidence of a life form generated from non-living material. This became in turn the starting point for the evolutionary process leading to all of the more complex living plant and animal organisms on earth. Natural selection is the mechanism in which favorable selection of random mutations succeed in the evolutionary process, while unfavorable ones fail. As such, nature's god of *natural selection* has replaced the God of nature and generated a theory analogous to that of "creative" evolution. Both require an untraceable process that adds just the right measure of complexity at just the right times while preserving all of the partially developed organisms from destruction. (Some may call this the definition of a miracle.) The intent being to prevent the natural flow of meandering evolutionary changes from wandering backwards or breaking down altogether as they would otherwise be inevitably destined to do.

There is no quantitative model or formulation even for the evolution of basic ingredients needed for a complex organism. Regarding the origin of merely a single component, that of human hemoglobin, computational scientist David Bailey writes,

> ... it is highly premature to pretend that anyone understands the process well enough to compute accurate probabilities.[32]

Traditional statistical analysis is considered inadequate and inappropriate since the basis for such computation is not available. Instead, naturalists often employ what is called "conditional" probability. With this method hypotheses may be retained and not rejected solely for lack of currently existing data that would otherwise be needed to support the likelihood of a theory. Evolutionary biologist Massimo Pigliucci in his book, *Making Sense of Evolution*, describes the approach,

> Evolutionary biologists need to move beyond physics...and past naïve falsificationism, to embrace the beautiful—if often frustrating—complexity of their discipline.[33]

The geological and biological observable data is for the most part qualitatively analyzed. And dating methods such as radiometric processes used in an attempt to quantify time scales from the past are not falsifiable. They rely upon assumptions about the initial residual levels of the elemental isotope in the material to be measured, for which there is no historical data. However, within the bounds of naturalism there is no available alternative to these theories, and therefore the popular paradigm persists. Since falsification of a theory extrapolated to the pre-historic past is not possible, it must rely upon such popularity and has actually done so quite well since its inception.

One difference is notable between the areas of study involving the theory of evolution and that of the origin of the universe with regard to the supporting observations or data. The astronomical observations and data upon which the theory for the origin of the universe is based are much more quantitative than the zoological or geological support for the theory of evolution. There is no precise quantitative description of the evolution of an organism. It is based on a random process which defies mathematical proof under the governing physical laws. One cannot, for example, construct a mathematical model for the evolution of a human being from a naturally existing group of chemical compounds in the same way that a mathematical model for the orbit of a planet is formulated. Referring to the popular paradigm of evolution, Victor Weisskopf, former president of the American Academy of Arts and Sciences, remarked that it was an "amazing '*fact*' that non-thinking 'Nature' has... generated intelligent beings and intelligible systems."[34]

Almost every theory of natural science has either repeatedly changed or been replaced over time. Ironically the paradigms of natural science are accepted and defended religiously during their reign of popularity. In the case of the origin of the universe, over the past century, paradigms last about 2-3 decades. Interestingly, this corresponds to some extent with generational changes within the scientific community as younger scientists revisit older theories. Sometimes because of the popularity of current paradigms or the waning historical knowledge of past ones, older theories are even revived as "new" theories.

The study of cosmology and the origin of life are primarily practiced by utilizing *inductive* conclusions, which may generally be multiple selective conclusions based on preferences, biases, etc. *Deductive* conclusions are rarely possible to employ in cosmology or the origin of life since they require a single logical conclusion uniquely resulting from verifiable observations. The conclusions sought in such investigations are intended to apply to times and places far outside the range of observation, and therefore defy deductive reasoning.

The origin of the universe lends itself to many possible theoretical scenarios, however flawed or incomplete, consistent with naturalism. This is to say that these possible naturalistic theories do not require intelligent design or the violation of current laws of nature or a significant change in the physical constants as well as other natural boundary conditions. On the other hand, the theory of evolution of living organisms based on random mutations has had no available options that are consistent with the restrictions of naturalism and its beliefs. For the purist it has been the only explanation in spite of its prominent "lumps". Lewis Thomas, former chancellor of the Sloan Kettering Cancer Center once remarked,

> Biology needs a better word than *error* for the driving force in evolution... I cannot make my peace with the randomness doctrine; I cannot abide the notion of purposelessness and blind chance in nature. And yet, I do not know what to put in its place for the quieting of my mind.[35]

Both theories of the origin of the universe as well as theories of evolution fall within the traditional definition of
pseudoscience in that they are not falsifiable. That is, they
are *faithfully* and *popularly* believed.

Pseudoscience Debate

By the mid to late 20th century the term pseudoscience was
no longer being applied to a broad range of investigation
which fell under the name of science including social and
behavioral studies. Falsifiability was no longer a requirement for a valid and recognized scientific methodology,
but traceability to naturalism was required. By this current
premise integrated medicine and parapsychology remain
in the category of pseudoscience. As these disciplines have
grown and become more widespread throughout universities and other institutions the debate has become more outspoken. Physicist, Sadri Hassani, of Illinois State University
calls pseudoscience a "societal mental disease too powerful
to be fought in the public arena".[36]

All the while, biblical theology along with its concept of
the world has remained an outcast to naturalistic science.
This became historically cemented since the mid 19th century and has become more and more the accepted opinion
with the practice of the modern paradigm philosophy in
natural science. Consequently the historical record of the
Bible is not considered to contain reliable data for the purpose of scientific study. Therefore any theory derived with
either a world concept based on biblical theology or one that
is supported by historical data from the Bible is considered

to be pseudoscience by those who adhere to a strict naturalistic approach.

Countering the growing popularity of naturalistic paradigms an ongoing exposition of historical biblical theology has been pursued regarding evolution and world origin. Early in the 20th century the biblical Text was expounded resulting in the construction of several theories of creation. One thing that these theories had in common was an accommodation of the naturalistic concept of an old earth and even older universe. But at the same time these were essential attributes upheld by naturalists to support the theory of evolution by random mutation. Within the past few decades since the 1980's, with the increased scientific data in astronomy and particle physics as well as genetics, further exposition of the biblical Text was explored within the context of various creation theories.

With this pursuit the intended outcome proceeded in two rather differing directions, both of which continue currently. One approach, building upon the theories of the early 20th century to accommodate an old earth and universe, included a modified form of evolution. Here divine intervention played a role to mitigate the unsettling notion that random mutation alone is capable of generating and advancing more complex organisms. These theological evolution theories took on many varied forms including "day-age" and "gap" theories, as well as "theistic evolution" and "progressive creation". The alternative theological approach to these theories adhered to the most literal and traditional interpretation of a young earth, which commonly became known as "creationism". With all of these came the temptation for a concerted effort to use and interpret, to their advantage, parts of the recent abundance of naturalistic data as evidence in various ways to support one or the other of these precepts.

Neither approach, either theological evolution or creationism, escaped being considered pseudoscience by the naturalistic community. But more importantly, both approaches hang precariously on the extrapolation of current scientific observations and data which are inherently changeable over time, making them subject to the same weaknesses as naturalistic theories. As the 21st century opened it found both contemporary scientists and biblical theologians pouring over data and theories or reading between the lines of the biblical Text respectively — all this in an effort to discover what may be undiscoverable and to comprehend what may be incomprehensible in *the world we know.*

4

Theological Caution

*Science cannot solve the ultimate mystery of nature. And
that is because... we ourselves are part of the mystery.*
— NOBEL PRIZE PHYSICIST MAX PLANCK

During the mid-1800's most scientific research was
carried on in universities. The German speaking city of Basel in the northwestern corner of
Switzerland was no exception. It was here that the family
of Friedrich Miescher lived. He was Professor of Pathologic
Anatomy at *The Basel University* and a practicing gynecologist as well. His son, Johann Friedrich Miescher, was
named after him. The boy was deeply influenced by his
maternal uncle, Wilhelm His, who was also a doctor. He
was just 13 years the senior of young Friedrich and was a
well known embryologist teaching Anatomy and Physiology
at the university. Wilhelm later became the inventor of the
microtome which is used to prepare thin tissues for examination under a microscope.

Friedrich was a shy and introvert child, but developed an early interest in science having been exposed to many books and intellectual conversation around the home. Fritz, as he was called by family and friends, spent much time alone just thinking. He was very intelligent and performed well in school. Living in a family of doctors, all those who knew him assumed that he would follow in this profession. But Friedrich had other ideas. He wanted to become a priest. However, family influences overshadowed this decision and he eventually pursued a career in medicine.

Friedrich completed his medical studies in 1868 by the age of 23. He struggled with the choice of which specialty of medicine to now pursue. At an early age he had contracted Typhus which left him with a hearing handicap. The work of a doctor often required the use of a stethoscope and he believed himself to be very inadequate without this ability. Remembering some of the encouraging words from his uncle, *"the ultimate questions about the development of tissues can be solved only by way of chemistry"*, coupled with his own interest in research, Friedrich decided to set out in chemical research. As time would tell, this turn of events in his life would prove to be most providential.

The whole concept of training for research was a bit new in the mid-nineteenth century. But Friedrich gained valuable experience and soon joined the prodigious laboratory of Felix Hoppe-Seyler. This was one of the first biochemical laboratories worldwide. It was located above the Neckar river valley in historic *Tübingen Castle*. Working out of a converted laundry room and kitchen, Friedrich was delighted to begin his experiments. He wanted to investigate the chemistry of cells and Hoppe-Seyer directed him to the white blood cell. Needing an ample supply of cells he soon found that the material he required was readily available at the local hospital. He gathered up the used bandages filled with blood and pus left from the treating of wounded soldiers there as a result of the Crimean War.

Back in his laboratory Friedrich had none of the modern methods of chemical analysis at his disposal. With patient determination, by repeated trials, he was able to develop a process for separating the "proteins" from the cell. He discovered that one of them did not behave as a protein at all. Further study revealed that this strange "nucleic acid" was not composed of certain materials common to proteins, making it particularly distinctive. The new substance which had come out of the nucleus of the cell he called "nuclein". Later it would be recognized as DNA (deoxyribonucleic acid). But it would be a full two years before this world shattering discovery was published in 1871 in *Medicinisch-Chemische Untersuchungen* (Medical-Chemical Investigations) and without the attention of headlines or news reels.

Friedrich believed that "nuclein" could be the key to the coding of hereditary information in the cell. He died from Tuberculosis in 1895 and his ideas about the role of nuclein were all but lost for 75 years from the time of its discovery. Until the 1940's proteins were still considered to be the most likely means by which genetic information was coded. In 1953 James Watson and Frances Crick, building on the collective and insightful observations or analysis of others who were studying the DNA molecule at the time, proposed a molecular structure for the DNA molecule which turned out to be the right one. With 20th century fanfare the now well known "double helix" structure was popularized. And within a few more years, nearly 100 years after the actual discovery of DNA by Friedrich Miescher, the code was finally broken in the mid-1960's leading to the rapid advancement of molecular genetics.

Erwin Chargaff was a biochemist, one of those more than a little responsible in paving the way for the determination of the DNA molecular structure in the early 1950's. He once reflected upon the long delay in failing to recognize the work of Miescher and attributed it to his being "one of the quiet in the land".[37] Only God knows why Friedrich was deterred from a theological career, stricken with Typhus, suffered a hearing affliction, diverted from practicing as a physician, to finally find himself under the mentorship of a biochemist studying bloody pus cells. This is no accident to the astute theologian. It is perhaps merely God's timing, lest keepers of 20th century, or even now 21st century, science become too proud and give a bit too much credit to themselves. But theologians would do well to consider the lesson with careful contemplation as well.

Scientific Conclusions Forever Changing
Scientific data and conclusions are very fluid, frequently changing and often completely reversing themselves as new observations and discoveries are made. Furthermore, scientific observations and data, when extended far into the past,

are limited by naturalism itself. They are only as good as one's faith in naturalism. Well known science writer Nigel Calder remarked,

> The natural pattern ... is provided by the cryptic unity of nature itself (belief in which is the chief act of faith of the scientist)...[38]

So to apply scientific observations and data in an exposition of the biblical Text, particularly when related to the origin of living things or the universe is always subject to regretful change or error.

It is possible, however, to add very valuable insights from science when considered in the appropriate context. First one should be very careful to take note of what is not really known and avoid overstepping these bounds on the fragile bridge of naturalistic observation or data. The biblical Text devotes less than 3 chapters or 80 verses to the entire events of the origin of the world and only 11 chapters for *all* of world history up to about 2000 BC. This represents a very small fraction of the 1189 chapters and over 30,000 verses in the Bible. Keeping this in perspective, there is far more that is not known than is known regarding the early world. Most of our theological knowledge and critical conclusions on this subject in the Bible hinge on just a few words or phrases.

Even a "literal" interpretation of the Text depends upon word meanings, limitation in language, context, and culture. So in addition to the information that is not given in the Text there are many uncertainties or unknowns that offer a challenge to a complete understanding of the brief information which is given. Modern science has added to the list of unknowns thus pressing the limits of naturalism even further. Strangely, some of these unknowns may provide interesting guidance in recognizing a depth of meaning beyond the surface. In doing so they may even illuminate a broader understanding of how the observable elements of the natural world actually mesh with the theologically defined world.

For example, the uncertainty principle in quantum theory posits that it is not possible to know both the exact location and movement of matter at the same time under certain conditions. It may only be determined to within a probability of knowing. This suggests that there is a limit to the role of determinism in the natural world and that what we call probability, for lack of a better understanding, may be something more. It may provide an avenue if not actual insight into the enigma of the *free will* of human beings within a natural world governed by the laws of nature. What we can only imagine as uncertainty in nature may be the very point where the supernatural actions of free will begin or we might say, God's interaction with the natural world. But for the naturalist randomness is a readily sanctioned alternative to divine interaction. Randomness is therefore accepted in naturalistic theories to some extent, although it has historically caused considerable consternation. Carl Jung, the founder of analytical psychology, once said, "I simply believe that some part of the human Self or Soul is not subject to the laws of space and time."[39]

Nevertheless, randomness is far from a recognized trait in the theological concept of the world. Biblical theology sees the uncertainties in the natural world under the control of God — *Uncertainties in the natural world may be uncertainties of nature to man but they are certainties of purpose to God.* From the words of wisdom spoken by King Solomon in the Bible,

> The lot is cast into the lap,
> But its every decision is from the LORD.
> *(Proverbs 16:33)*

There is an old rabbinical saying, "Coincidence is not a kosher word!"

Another point of interest in quantum theory is the concept of "entanglement". It asserts that groups of individual particles have certain characteristics which depend upon the characteristics of others in the group even if they are completely isolated from each other, that is, beyond communication by the speed of light. This suggests an interaction which is not explainable with naturalistic reasoning. And it further implies possible limits to human reasoning with regard to understanding the workings of the world. Once again, the theological concept of the world based upon the Bible would expect such deviations from the natural world and its laws. It should also not be a surprise to the student of biblical theology to find that perplexities beyond human reasoning or conception may be uncovered at the leading edge of modern scientific investigations. From the Text, The prophet Isaiah proclaimed God's words,

> For as the heavens are higher than the earth,
> So are My ways higher than your ways
> And My thoughts than your thoughts.
> *(Isaiah 55:9)*

The age of the earth has become in modern times a great point of disparity between traditional biblical theology and the modern paradigms of science. Additionally, a considerable debate has even emerged among the differing theological views. The development of the theory of relativity by Albert Einstein in the early 20th century established a new perspective when considering the passage of time. And a further development by Einstein included the conditions imposed by gravity and was called *general relativity*. With the assumption that the speed of light is a constant, the primary premise is that time becomes a variable which depends upon the frame of reference. So the passage of time is different for different frames of reference. Gravity complicates the result causing variations in time and space itself. Therefore, in the extreme situation of the origin of the earth and the universe, the accounting of time may have been much different than is perceived by us today. The result would depend upon either the frame of reference or the proper analysis with the applicable physical constants under the extreme physical conditions of the distant past. It is conceivable that both a long time as well as a short time may actually represent the events associated with the origin of the world depending upon the frame of reference and existing conditions.[40] As the Text reflects God's perspective,

For a thousand years in Your sight
Are like yesterday when it passes by,
Or as a watch in the night.

(Psalms 90:4)

The concept of time itself remains a particularly difficult enigma for the contemporary scientist. The traditional view of absolute time was destroyed by the theory of relativity which established the premise that time is relative to the observer. But the scientific study of the universe is based on cosmological time. As with time expressed in biblical theology or in biological time, there is a preferred reference as well as a preferred direction in cosmological time. It proceeds smoothly and irreversibly from the past to the present. And so without explanation or understanding, cosmological models depend completely on this peculiarity of time.

Dueling Apologetics

Commenting on the ever growing number of speculations for the demise of the dinosaurs, physicist Paul Huffman wrote,

> The list of suspects is a long one, and it is a tribute to what human ingenuity can come up with when confronted with something it can't understand.[41]

Many theories of theological evolution followed in the wake of naturalistic evolution and world origin paradigms. These

theories often grew both in number and popularity in pro-
portion to the rise in credibility given to these paradigms
by scientific naturalists. Although there is much variation
in the theories, among the characteristics they all have in
common is that they accommodate some form of evolutionary process for the living organisms and a long period
of time for this process to play out, similar to the naturalistic paradigm. This generally results in multiple theological
conflicts with long standing biblical principles which are
threaded throughout the Text. And for the most part these
are very difficult to resolve. Even outspoken atheist Richard
Dawkins said in a PBS interview,

> For one thing, if I were God wanting to make a
> human being, I would do it by a more direct way
> rather than by evolution.[42]

Biblical theology may be described by *doctrines* which
represent the understanding of a subject or principle based
upon the Text as a whole. There is no particular limit on the
number of bible doctrines with a hundred or more sometimes identified. But there are a few which are primary to
the central theme of the Bible. These include the subjects of
God, Jesus Christ, Man, Sin, Redemption, and the Future.
There is also a doctrine regarding the Bible itself and its
divine authority. The primary doctrines have a long history
as old as the Text, from its early Jewish heritage through first
century Christianity. The principles of biblical doctrine are
based on a strict interpretation of the Text, which is both
cohesive and self consistent. All parts of the whole support

the central thematic principles from which historical doctrines are related. Many of the attempts to accommodate naturalistic paradigms which emerge from contemporary science lead to doctrinal conflicts. As a result of such complications, the theologian is tempted to loosen the restraint on biblical cohesiveness, lessen the weight of historically key biblical references, and redefine doctrinal outcomes.

As an example of the theological complications, consider the doctrine of man and sin. God's original complete creation of the world including living things and human beings was recognized as "good" (Genesis 1:31). Perfection is a long standing attribute of God, supported by the Text and recognized by both the ancient Jews as well as Christians. But evolution, even theological evolution, is premised upon a process that is anything but perfect in its ongoing building up and tearing down as it progresses toward the completion of the creation process. Death and decay are a prevalent outcome all along the way. This conflicts with the biblical principle that death is the result of sin, and that decay is its inevitable consequence. This principle, first introduced in Genesis, Chapter 3, understood by the ancient Jews, and reflected upon over and over in the Text, considers that sin began in human beings with Adam. Because of his transgression, the world would no longer be the same, with death and decay thereafter holding sway, only to be redeemed with God's intervention at some time yet to come in the future.

The Bible draws a clear distinction between the nature of man and that of other living creatures. Human beings are characterized with a soul. The Bible clearly assumes that the soul is eternal in existence (e.g. Matthew 10:28).

This uniqueness creates quite a problem within a common evolutionary process that produced all of the living organisms as well as human beings. The theological evolution theories are forced to awkwardly conceive of pre-human creatures which have evolved, being otherwise human but not endowed with a soul. These pre-human creatures could actually be co-existent with Adam and his descendents and would include Adam's parents.

Furthermore, since naturalism and evolution, taken to their logical conclusion, have no place for the concept of a human eternal soul, they are in conflict not only with the past but with the future. Human beings are tied inextricably to the naturalistic universe in an evolutionary progression which will eventually lead to their extinction, one by one, as they die. All of this is at odds with the biblical doctrine of man as well as the doctrine of redemption.

An obvious consequence of naturalistic thinking is that we are not necessarily alone in the universe. If life by natural processes could appear on earth, than why could it not appear in some other place and some other time? The *cosmological principle* upon which the physical theories for the origin of the universe are founded states that the matter in the universe is homogeneous and isotropic (uniform) everywhere if taken on a large enough cosmic scale. All of the cosmological mathematical formulations depend upon this principle. In other words, it is assumed that the planet earth and its inhabitants hold no special place within the universe.

So it is only "natural" to consider that there may be other earth-like planets with life on them, and that there may well be other intelligent life. Within contemporary science this is no small matter. Nearly every major astronomical research facility has some activity involved in the search for life elsewhere in the universe and in particular intelligent life. To this end radio telescopes and other instrumentation are constantly scanning the skies for a bit of radiation which may suggest that it was transmitted from a population of intelligent beings far away in space and time.

Biblical theology is clearly *God centered* but it is also very much *earth focused*. Regardless of whether the earth is physically at the center of the universe, it is not hard to see that the biblical Text associates the universe and its creation with man on earth. Now, one could conclude that the concept of life elsewhere, created by God, as part of the as yet undiscovered "dominion" or rule of man over all living things is consistent with this principle as described in biblical Text of Genesis 1:26.

> Then God said, "Let Us make man in Our image, according to Our likeness; and let them rule over the fish of the sea and over the birds of the sky and over the cattle and over all the earth, and over every creeping thing that creeps on the earth."

However, the Text directs this rule to that which is on the "earth". So living things, out of the reach or influence of man on earth, would be extraneous to this theme and therefore outside the realm of biblical theology. Furthermore, God's relationship with man and His redemptive plan for man outlined in biblical theology is inextricably connected to the life of Jesus. The life of Jesus is unique in biblical theology and His purpose in the world, on earth, was historically connected to earth. The life, death and resurrection of Jesus as a sacrificial act was earth focused and experienced on earth in order that man on earth would know and respond. In biblical theology, intelligent life (aside from angelic beings) is associated with humans as they are made in God's "image". The future as well as the past is earth focused in biblical theology. The eternal home of those who believe and accept God's redemptive offer of salvation from this decaying earth is a "new earth" (Revelation 21:1). This new earth is described as a new creation on earth which unlike the old, will not decay. Again this is earth focused.

Biblical theology describes God as uniquely interested in human beings and it describes the essential need for human beings to have a relationship with God. This relationship is to be eternal, beyond the limits of the natural universe. If there were human beings elsewhere in the universe, they would be distanced from all the unique history and future expectation of biblical theology. Considering the uniqueness of God's attention to earth, it would suppose either a necessary mission endeavor of cosmic proportions on the part of humans here on earth to reach the far flung human neighbors, or that Jesus would have duplicated his life and ministry multiple times, including his sacrificial death and resurrection in order to make an appearance on other worlds. Neither option would render itself at all consistent with biblical theology.

Francis Collins, former Director of the Human Genome Project, was an atheist throughout his young adult years as a student and medical doctor. But while watching his patients in life and death issues he began to ask questions such as, "What is the meaning of life?" "Why am I here?" "Why does mathematics work, anyway?" "If the universe had a beginning, who created it?" "Why are the physical constants in the universe so finely tuned to allow the possibility of complex life forms?" "Why do humans have a moral sense?" "What happens after we die?" He had been taught to reason as a scientist and his reasoning mind sought answers. He found answers through the logic in the writings of C. S. Lewis and became a believer in Jesus Christ, realizing that ultimately faith is the answer. He led the renowned research program to read out the 3.1 billion letters of the human genome which he calls our instruction book, a molecule written in God's language. The more he learned about the creation, the more he saw God in it.

Because of his high profile as the Director of the National Institute of Health, he has had many occasions to speak of his faith and belief in God. In his book, *The Language of God*[43], Collins outlines arguments for theistic evolution in which God is the initial creator but plays no intermediate role in the creation as it evolves. He accepts as open and unresolved the issues of the origin of life as well as the uniqueness of man, which is considered to defy evolutionary explanations. The divinity of Jesus Christ and other theological precepts from the Bible are generally not addressed. So it may be seen that while theistic evolution could be compatible with the "science" of origins it is often the "theology" of theistic evolution which has a difficult time harmonizing with the Bible.

Sometimes the tendency of theological apologetics when confronted with naturalistic science is to disproportionately accommodate the popular paradigms as if compelled to do so. Scientific paradigms are based on changeable theories. But what should be more disconcerting is the limited and restricted definition of the world within the framework of naturalism.

For example, it is recognized from scientific inquiry of the present natural world that "time" is a variable along with space itself, and that it is dependent upon the reference frame and other physical factors. Then upon reflection, from this insight one may gain a greater appreciation for the incomprehensible marvel of how God created the world. Further one could grasp how limited is man's understanding of something as simple as a period of time, such as a "day" from "evening to morning" (Genesis 1:5). While the few verses in the Text can remain fully accurate much is revealed about that which is not said. After all, from God's perspective, the entire theory of general relativity may well be hidden in those few choice Hebrew words. It is one thing to realize from the study of science that the Hebrew word *yom*, generally translated as "day", may have a much deeper meaning than our current understanding while actually retaining its literal meaning of a 24 hour period as well. But it is another thing altogether to surrender the accuracy of the Text to accommodate and reconcile a naturalistic theory of origins which is unarguably premised upon an exclusion and denial of the entire spiritual realm.

Even among those theologians who hold to the accuracy of the Text it is very risky to build an apologetic argument in the attempt to fill in or expound too much upon what is not said in the Text based upon scientific observations or data. Such arguments often require retraction as new scientific discoveries emerge, past evidence is discredited, or conclusions change.

An Example to Consider

Isaac Newton has been respected as perhaps the greatest of all scientists. And yet based upon his writings and those who knew him he was actually more of a theologian than he was a scientist. He revered the Bible and believed it to incorporate hidden information for the careful student. Rare papers, revealed in recent times, containing his written notes showed a prediction that he made for the end of the world based on the book of Daniel. He suggested that the end would come sometime after AD 2060. But he cautiously qualified the prediction with this comment,

> This I mention not to assert when the time of the end shall be, but to put a stop to the rash conjectures of fancifull men who are frequently predicting the time of the end, & by doing so bring the sacred prophesies into discredit as often as their predictions fail.[44]

Newton was also just as reserved regarding speculations of the past as he was with these predictions of the future. In all of his life he never showed any doubt concerning the authority or accuracy of scripture, and he recognized the Mosaic authorship of Genesis. But he was careful to limit what may be read into or inferred from the description of creation,

> As to Moses, I do not think his description of [the] creation either philosophical or feigned, but that he described realities in a language artificially adapted to [the] sense of [the common person].[45]

He believed that the day by day outline and events of creation were written to be best understood and appreciated by the people of his time as well as all others who may read it. Although consistent with the actual events, it was simplified and it adopted language and terminology which was within the common vocabulary and comprehension. It was completely accurate. But for Moses "To describe them distinctly as they were in themselves, would have made [the] narration tedious and confused...and become a philosopher [scientist] more than a prophet."[46]

Therefore, Newton did not place a great emphasis on an exaggerated exegesis of the words or phrases, believing them not to be intended to convey such a detailed or analytic image. While at the same time he believed the description to be fully knowledgeable and no way in error. He never considered that his own or other's advancements in natural philosophy or science would ever be at odds with the biblical description of creation whether in his lifetime or hundreds of years later. Because of his biblical world view he saw no need to strain the Text over such things as the position of lights in the heavens or the events of a day. He took these lights to be "their apparent not real place". And he considered that the stars as distinctive bodies were made on some unknown day but described and shown on the 4th day when they became apparent "at such a time, as they were made such phenomena".

On the other hand, Newton's writings are replete with many integral references between the natural world and God's creative initiative as well as His continual control. Therefore to him, theories of creation including evolution which are precipitated from a naturalistic world view would be completely inconsistent. Although Newton explored the entire realm of nature, both as a philosopher and scientist, his theological foundations undergirded his efforts. His work led to a timeless collection of scientific and mathematical accomplishments known as classical or Newtonian mechanics along with classical optics, which are revered and utilized even today. It remains scientifically relevant over 300 years later. I think it no accident that such success followed a God centered world view inextricably combined with theological wisdom and discretion.

Modern cosmologist Robert Sanders opined,

> ... any discussion of philosophy is to enter the realm of dangerous speculation. In general this is a risky business for a scientist. On the other hand, it is difficult to avoid philosophy in any discussion of cosmology.[47]

He wrote that the idea of a purposeless universe is ... "essentially a modern assumption that cannot be proven in a purely scientific context."[48]

If a description of the origin of the universe were formulated based on the purposeless philosophy of contemporary science it might read something like this — *By faith it is believed that the universe was created by itself a very long time ago by unverifiable naturalistic processes and evolved over that time to the present day state by some type of natural physical laws.*

By comparison, this summation of a naturalistic origin is about the same level of detail regarding the creation of the universe as that given in the first few verses of the Bible with a divine origin. To say much more about the details of how the universe was created based on scientific analysis, or for that matter, by expository biblical theology as well is getting into the realm of speculation.

Imagine being given a snapshot of a person taken against the backdrop of a non-descript location and asked to determine solely from the photograph — with no other additional information — the entire life history of the individual, including the exact time to the minute and second of their birth. But how infinitely easier this would be than trying to determine the life history of the universe from our "snapshot" of the earth and outer space in the present age. At least in the case of the person in the picture we have other relevant knowledge such as our own experiences in times past, verifiable experiential knowledge of other people and lives, as well as documented historical knowledge of similar past times and places. With the history of the universe we have none of that (excluding biblical references). We do not even know if the force of gravity or the laws of nature have always been the same.

The Bible does add a few more historical insights about the early universe in the first two chapters. And the entire remaining Text certainly reiterates and supports many of the elements of the brief description found there although adding little more detail. But what it lacks in detail only suggests that there is much which is beyond speculation. One would think that as knowledge of the complexity and intricacy of the natural universe increased through contemporary science that we would appreciate all the more the shortcomings of our human comprehension. What is so concisely and succinctly presented in the Text leaves insatiable wonder about *the world we know.*

5

Perspective from the Unknown

What we know is a drop, what we don't know is an ocean.
— ISAAC NEWTON

Some years ago when the debate over the new discoveries in astronomy was beginning to flourish, a well known scientist, Robert Jastrow, delivered a presentation to his fellow scientists in which he said,

> For the scientist who has lived by his faith in the power of reason, the story ends like a bad dream. He has scaled the mountains of ignorance, he is about to conquer the highest peak; as he pulls himself over the final rock, he is greeted by a band of theologians who have been sitting there for centuries.[49]

Anaximander was a pre-Socratic Greek philosopher of the 6th century BC. He is often credited as the first to theorize evolution as a process by which living organisms as well as man emerged. His writings were based solely on his faith and beliefs. They contained no scientific reasoning. He was considered a pantheist, associating and elevating nature to the status of a god. Modern evolutionary theories are much more recent in origin and are found in the published works of the French nationalist, Jean-Baptiste Lamark between 1802-1822. His writings would later influence Charles Darwin. Contemporary naturalist Richard Dawkins has been a leading voice for these evolutionary theories. He once degradingly described the religion of Pantheism as "sexed-up atheism"[50]. But the gods of the naturalist have historically been the laws of nature themselves. The definition given by Merriam Webster for pantheism is "a doctrine that equates God with the forces and laws of the universe"[51]. So ironically, for the atheistic naturalist, it would appear that atheism may just be pantheism in denial.

For the naturalist without God, their god is ultimately the Law of Nature. Not just the laws of nature — for the naturalist is forever further seeking the great god of the Law of Nature — a single natural law which encompasses all of the laws of nature into one. In the quest for this ultimate goal, theories often defined as the "grand unified theory" have become popular. Within contemporary science since the 1970's, these theories attempt to combine at least three of the four forces of nature into a single model, although there is currently no solid evidence that nature is described by such a "theory of everything". And the force of gravity remains a stubborn and unwilling holdout, defying nearly a hundred years of attempted scientific collusion to this end. Within contemporary science it has given rise to reductionism taken to the extreme.

If such a grand Law of nature were found, it would without doubt be worshiped by many, made of matter and energy, not so unlike the idols familiar to the biblical theologian, which are made of wood or stone. It would, from a theological perspective, perhaps be reminiscent of the building of the Tower of Babel, reaching to the edge of the universe as it were, with telescopes as their ladders.

Consequences of the Controversy

For the naturalist their story of the origin of man and the universe was a conscious construction. It was decided that the creation was simply the beginning of a process which has played out as a consequence of certain rules, observable within the physical environment in which we dwell. These rules of the physical world have always been the same and have always existed. By *faith* this is so. Nothing else is necessary and nothing else exists beyond the accepted realm of nature observed by the senses. Modern science has been very proficient in fueling technologies of the future. But it has been fraught with shortcomings in uncovering the mysteries of the prehistoric world. Technology is a well accepted fruit of modern science. Likewise cosmological theories are another fruit of modern science, but such theories are, both present and past, forever shrouded in irresolvable controversy.

So one might ask why there is such a concerted effort of investigation within naturalism to establish a story about the past, even if it is burdened with unexplainable difficulties which challenge the credibility of scientific methodology. A common reflection of many naturalists on this question is the inborn desire to know where they came from. But in reality they have already decided where they came from. A more reasonable explanation is that having set the bounds and restrictions for the physical world and its reality, it is important for naturalism to dispel all other options. This means building a case for the natural origin of man and the universe, however precarious and however much faith is required. To be sure, any proposition explaining the distant past, whether natural or supernatural is inherently and "religiously" faith-based.

The impact of naturalism has been great over the centuries. The first mention of it in the Bible long ago is found in Psalms 14:1,

> The fool has said in his heart, "There is no God"...

In more recent times the culture has embraced naturalism and generated through the educational process its popular and exclusive acceptance. Although perhaps not the intent of its authors, it became the basis of social movements such as those led by Margaret Sanger and Carl Marx as well as the societal influence of Nietzschean racism, Freudian amoralism, and national movements of military imperialism. Referring to naturalistic evolution, historian David Jorafsky wrote,

> ... an historian can hardly fail to agree that Marx's claim to give scientific guidance to those who would transform society has been one of the chief reasons for his doctrine's enormous influence. [52]

On reflection, it would be hard to enumerate any benefits or contributions to mankind that are the result of a purely naturalistic concept of the world or of its origin. The primary consequence of naturalism and the exclusion of God is the lowering of human beings to the level of all other living creatures. Human lives are of no greater value or significance. Their lives end with physical death just as all other animals. Since the traits of human beings are just a matter of genetics, then the manipulation and control of their physical makeup and offspring is of no consequence. The laws for human beings are no different than the laws which apply to the animal kingdom as a whole.

The Bible prophet Daniel envisioned a time far in the future when mankind's earthly abode was nearing its end and "... knowledge will increase" (Daniel 12:4) but wisdom of God would be lacking as "many will go back and forth". Science writer and devout atheist, Isaac Asimov, once remarked, "... science gathers knowledge faster than society gathers wisdom."[53] He was probably referring to nuclear weapons, but how ironic that it is actually a more apt description of what Daniel foresaw as scientific knowledge when used misguidedly to undermine God and diminish His own creation. Such a caution is found in the Text,

> O Timothy, guard what has been entrusted to you, avoiding worldly and empty chatter and the opposing arguments of what is falsely called "knowledge"—
> *(1 Timothy 6:20)*

With the foundations laid and the doors to contemporary science just beginning to open in the later part of the 19th century, theologian W. A. Snively commented,

> There is no more vital and anxious thought in the religious life of to-day than the supposed conflict between science and religion... And so long as scientists attempt to teach theology, and theologians insist upon refuting what they choose to dignify by the name of science, so long there will be a terrible warfare of words; but it will not touch nor jeopardize for a moment the indestructible harmony between true science and true religion, between a right reason and a devout faith, between the broad page of nature, written by His own finger through the long processes of His own law, and the page of inspiration, written by the human amanuensis of His own Spirit. There is one point, however, in the universe, in which nature and revelation meet; one point in which the visible creation comes

in contact with the invisible and supernatural forces which pervade the universe. That solitary point is the incarnation of the Son of God [Jesus Christ]. In it nature and revelation mysteriously meet and harmonize; as by it this human nature of ours — the very crown and glory of the visible creation — is taken into union with God. Here the ultimate mystery of science and religion meet and harmonize and are at one; as by the incarnation the nature of man is allied to the throne of God in a union which can never be divorced, and which waits for its final epiphany for the manifestation of the sons of God.[54]

God made possible for that faith to be real and filled with hope through Jesus Christ, as he bridged the gap between man and God, between the natural and the spiritual world. His sacrifice not only made it possible to understand the two but to be a part of both forever, through faith and trust in Him to redeem degenerate man from this natural world .

Biblical theology will always stand on its own if allowed to do so. Volumes have been written based on topics related to the nature of the world and its origin for which the underlying percepts are either totally unknown or for which there is very limited understanding. This occurs both from the perspective of a theological world concept as well as a naturalistic one. One advantage of the theological approach based on the Bible is that the various anomalies in nature should not disturb it. Such anomalies are only to be expected, considering the existence of the spiritual world and its connection to the natural world. But the insights into the interpretation of scientific observations and data may lead to much greater understanding. This is so when the temptation is resisted to presume or extrapolate with undue confidence beyond that which is known as do some of the popular paradigms. Many of the great scientists of the past, including Isaac Newton, were theologians as well as scientists. They rarely separated the two in their work and it helped them to see beyond the obvious. Naturalism will always be limited to human reasoning.

Based on the theory of general relativity it is not possible to "see" beyond a so called "event horizon" which is determined by the expansion of the universe and the speed of light. This limit actually defines the edge of the visible universe itself. Beyond the natural horizon, beyond the boundaries that naturalism has placed on itself by human reasoning, is a world greater than the natural world. Biblical theology describes this larger world and speaks of its origin in just a relatively few verses. And in those few verses resides the totality of what God chose there to explain. Both the theologian as well as the scientist should be careful not to expound beyond what is actually known. With the proper perspective scientific observations may provide valuable insights to a world of both the spiritual as well as the natural. But only with faith can we apprehend the world beyond the senses. In the words of Helen Keller who understood this well,

> ... faith ... has made my limitations ineffectual if
> not trivial ... Through faith alone can I fulfill the

two senses I lack — sight and hearing ... Faith has
the ingenuity to bring me insight, and I know
where I am going ... [55]

Clear Vision

Biblical theology presents a description and understanding
of the universe and its history over a vast scope far beyond
that of natural science. Interwoven with this description
is also a purpose as with everything designed. And what's
more, the designer is introduced. The universe we live in is
not just a stage. It is the record of this grand plan. The universe contains not only the beauty of His perfection within
its creation, but it is also a reminder of man's deliberate
defiance. With that the consequences of sin forever marred
the landscape of the earth and even the far reaches of the
starlit universe. The Text describes it as a world in need of
restoration.

We know that the whole creation has been
groaning ... right up to the present time.
(Romans 8:22)

One can admire the vibrant color and pleasing shape
of a flower and yet all too soon the petals begin to fall. A
majestic oak tree stands strong and yet its branches may
crack and decay. A tall mesa in the desert reflects the light
with pleasing shadows and shapes, but contains the story
of erosion and violence in its steep scarred walls. Viewing
the stars is often mesmerizing as we stare captivated by

their silent brilliance. Even the stars may flicker and fade predicting the eventual future of this world. What could be more an example of God's creative ability than is the intricately and wonderfully made human mind and body. Adam's body was designed to last forever, but now only "three score and ten".

Natural science helps us see both the beauty and perfection of the creation as well as the inevitable course of decay and conflict. However, to obtain a clear and unobstructed view, it is very advantageous to peer through the lens of biblical theology. Other spectacles may be deceivingly myopic. Ironic it is that the first telescope in space was named after Edwin Hubble, an agnostic, whose views on the cosmos were shaded by his preconception of an infinite universe. And this grand telescope, a marvel of modern technology, proved to be short sighted due to a defect in the optical design. The fix required a corrective lens to be shuttled up to it in space and then later attached. It truly needed spectacles to see clearly.

Werner Heisenberg once said,

> The existing scientific concepts cover always only a very limited part of reality, and the other part that has not yet been understood is infinite.

The realm of science is a subset of the realm of biblical theology. Science falls within the confines of the natural world as well as in the timeframe of the present. For it to venture into the past or future exceeds the bounds of the scientific method. Certainty of conclusions is then forfeited. Biblical theology encompasses both the natural and spiritual world, past, present, and future. The key source of knowledge is the Text. Since the Text does not provide many of the details, at least within human understanding, extrapolation using the tools of science may forfeit the certainty of any conclusions as well. Science allows the theologian to discover and view more of God's design in the present natural world, while biblical theology can offer insights into the understanding of the natural world. But biblical theology can also provide guidance and appreciation for a larger world which although natural to God, is beyond the nature of humans. As humans, we are confined in time and nature, as is the science we are fit to practice.

Physical science hangs on the requirement and belief that the laws of nature which are observed in the present are also fully valid in the future just as they are assumed to have been valid at any point in the past. Predictions into the future based on the expected life cycle changes in the sun suggest the earth will be uninhabitable within about 1 billion years, give or take a few hundred million. This will occur as the temperature of the sun increases causing the oceans to undergo runaway evaporation and loss of CO_2 in the atmosphere, necessary for photosynthesis. But contemporary science has a long list of other less predictable agents of doom for the earth. These include impacts by comets or asteroids, a massive stellar explosion or supernova, irregularities in the earth's orbit or its rotational axis, and a highly virulent disease. And then, of course there are anthropologic predictions such as nuclear holocaust, resource depletion, or human-induced climate change.

The naturalist believes that contemporary science using the tools of technology along with an understanding of natural laws can save itself and prescribe its own future. To this end contemporary science tracks asteroids, studies population growth, and models global temperatures, looking for effects caused by fuel emissions. Long term motivation for space exploration is seen by some as a possible escape plan from an exhausted planet earth to Mars or elsewhere.

Biblical theology illuminates, by contrast, contemporary science's adoption by naturalists of a belief in nature and its physical laws as its god and savior. Human beings, as part of nature and created by nature, are to be the enablers of the salvation for the inhabitants of the earth. So it should be no surprise that future doomsday threats to the earth are a very significant activity of contemporary science. And anthropologic threats are of particular priority, being worth any sacrifice.

The Bible just as assuredly predicts the doom of this world and furthermore, all of nature itself. Hidden in the biblical description of the creation of the world is a portent of its outcome in the aftermath of sin (Genesis 3:15). The "seed" of the woman looks forward to Jesus Christ, the promised deliverer and God's plan of redemption for the eternal soul of man. With His death on a Roman cross at Calvary and resurrection from the grave Satan was defeated, the curse of sin was lifted, and the salvation of all who would believe in the Savior was fully accomplished (Galatians 3:13-16). Nonetheless, the natural world is already destined for its own eventual end as modern observations of natural science commonly suggest, by "intense heat". A similar conclusion had already been recorded for centuries and is well described in the Text including the future demise of the world.

> ...the heavens will pass away with a roar and the elements will be destroyed with intense heat, and the earth and its works will be burned up.
> *(2Peter 3:10)*

Additional details are given in the biblical narration which unfolds in Revelation. The end of the earth is described like nothing ever before. It will truly be "climate change" alright — on Divine proportions. What naturalists attribute to random interstellar events, or more locally to the consequences of carbon emissions from burning of fossil fuels, has another explanation here. For biblical theology the end of the earth is the expected conclusion to a world in decay, under God's complete control. Overpopulation, resource depletion, and global environmental impacts are not so much of a concern for biblical theology, considering that it is not nature but the God who created nature who already has the schedule planned. And just as the first pages of this world story have been recorded, so are the last pages already penned as well.

All is connected together from the distant past to the inevitable future of the world both in substance and in purpose. Biblical theology achieves its greatest understanding when sustained by the greatest faith. The world and its purpose illuminated by its creator may then be revealed. Likewise, contemporary science at its best reaches its most beneficial accomplishments when focusing on the natural world of the present in which we live. The world and its resources may then fulfill God's intended provision for people, whom He loves, in *the world we know.*

Notes

Introduction

1   Quoted in: Ferris, Timothy, *Coming of Age in the Milky Way*, New York, Morrow, 1988, p177.

2   McLeish, Tom, "Thinking Differently about Science and Religion", *Physics Today*, February 2018, p10.

Chapter 1: Nature's Mysteries

3   Tiner, J. H., *Isaac Newton: Inventor, Scientist and Teacher*. Milford, MI: Mott Media, 1975.

4   Newton, Sir Isaac, *The Mathematical Principles of Natural Philosophy*, Published by Daniel Adee, New York, 1846, p74.

5 "Why do measurements of the gravitational constant vary so much", from: http://phys.org/news/2015-04-gravitational-constant-vary.html

6 *The General Scholium* online, trans. Andrew Motte, 1729. From: https://isaac-newton.org/general-scholium/

7 Heisenberg, Werner, *Physics and Philosophy: The Revolution in Modern Science* (1958), Lectures delivered at University of St. Andrews, Scotland, Winter 1955-56.

8 Cham, Jorge and Whiteson, Daniel, *We Have No Idea*, Riverhead Books, New York, 2017, p2.

9 Cham and Whiteson, op.cit. p91.

10 Cham and Whiteson, op.cit. p289.

11 Cham and Whiteson, op.cit. p280.

12 Schrödinger, Erwin, *What is Life?*, Cambridge University Press, 1944, p76.

13 Lewes, C. H., *Problems of Life and Mind*, first series, J. R. Osgood and Company, Boston, Vol. 2, 1875, p369.

Chapter 2: Different Worlds

14 Flammarion, Camille , *L'atmosphère : météorologie populaire*, Paris, 1888, p163.

15 Joubert, Joseph, *Some of the "Thoughts" of Joseph Joubert*, published by William V Spencer, Boston, 1867, p32.

16 Zwicky, F. (1933), "Die Rotverschiebung von extragalaktischen Nebeln", *Helvetica Physica Acta*, vol. 6, pp110–127.

17 Calder, Nigel, *Violent Universe*, British Broadcasting Corp., 1969, p25.

18 *Writings of Henry David Thoreau*, edited by Bradford Torrey, Houghton Mifflin and Co., Boston and New York, 1906.

19 Dreyer, John Louis Emil, *The History of Planetary Systems from Thales to Kepler*, Cambridge University Press, 1906, pp20,37,38.

20 Paley, William, *Natural Theology or Evidences of the Existence and Attributes of the Deity*, R. Faulder, London, 1802, p1.

21 Von Braun, Wernher, In a letter to California State Board of Education, 14 September 1972.

22 Smith, GH., Atheism: *The Case Against God*,
    Prometheus Books, 2003, p155.

23 Dawkins, Richard, *The Blind Watchmaker*, Norton,
    New York, 1986, p317.

24 Asimov, Isaac, *Free Inquiry*, Spring 1982, p 9.

Chapter 3: Boundary of Science

25 "History of the Scientific Method", from:
    https://explorable.com/
    history-of-the-scientific-method

26 Capra, Fritjof: *The web of life: A new scientific
    understanding of living systems*, First Ed.,New York:
    HarperCollins, 1996, p6.

27 Sanders, Robert H., *Deconstructing Cosmology*,
    Cambridge University Press, Cambridge, UK,
    2016, p35.

28 Calder, Nigel, *Einstein's Universe*, British Broadcasting
    Corp., 1979, p120.

29 Sanders, op. cit. p3.

30 Kragh, Helge, *Cosmology and Controversy*, Princeton
    Universtiy Press, Princeton, New Jersey, 1996, p191.

31 Kaufman, Rachel, "Rebel Rebel", *Radiations*, American Institute of Physics, College Park, MD, Fall 2016, p18.

32 Bailey, David H, "Evolution and Probability", Report of National Center for Science Education, Vol. 20, No. 4, 2001, p2.

33 Pigliucci, Massimo and Kaplan, Jonathan, *Making Sense of Evolution*, University of Chicago Press, Chicago and London, 2006, p227.

34 Weisskopf, Victor, "The Frontiers and Limits of Science", *American Scientist*, 65, July- August 1977, p405, as quoted by Henry Morris in *The Biblical Basis for Modern Science*, Master Books, 2002, p19.

35 Thomas, Lewis, "On the uncertainty of Science", *Key Reporter*, vol. 46, Autumn 2000), p2.

36 Hassani, Sadri, "The Dangerous Growth of Pseudophysics", *Physics Today*, May 2016, p10.

Chapter 4: Theological Caution

37 *Rebels, Mavericks, and Heretics in Biology*, edited by Oren Harman, Yale University Press, New Haven and London, 2008, p.96.

38 Calder, op. cit.

39 Quoted in: Davies, Paul, *God and the New Physics*,
    Simon and Schuster, Inc., New York, 1983, p72.

40 See for example: Schoeder, Gerald, *Genesis and the
    Big Bang*, Bantam, 1991.

41 Hoffman, Paul, "Asteroid on Trial", *Science Digest*,
    June 1982, p62.

42 Dawkins, Richard, "Faith and Reason", interview
    transcript, Public Broadcasting Service, hosted by
    Margaret Wertheim, from:
    http://www.pbs.org/faithandreason/transcript/
    dawk-frame.html

43 Collins, Francis, *The Language of God: a Scientist
    Presents Evidence for Belief*, Simon & Schuster, 2006.

44 "Papers Show Isaac Newton's Religious Side, Predict
    Date of Apocalypse", *Associated Press.*, 19 June
    2007, from:
    https://web.archive.org/web/20070813033620/http:/
    www.christianpost.com/article/20070619/28049_
    Papers_Show_Isaac_Newton%27s_Religious_Side,_
    Predict_Date_of_Apocalypse.htm

45 Brewster, Sir David, *Memoirs of the Life, Writings, and
    discoveries of Sir Isaac Newton*, T. Constable and Co.,
    Edinburgh, Vol. 2, 1855, p450.

46 Ibid. p452.

47 Sanders, op. cit. p5.

48 Sanders, op. cit. p10.

Chapter 5: Perspective from the Unknown

49 Jastrow, Robert, *God and the Astronomers*, W.W. Norton and Co., New York, 2000.

50 Dawkins, Richard, *The God Delusion*, Bantam Books, New York, 2006, p40.

51 Merriam Webster Dictionary, from: https://www.merriam-webster.com/ dictionary/pantheism

52 Jorafsky, David: *Soviet Marxism and Natural Science* Columbia University Press, New York, 1961, p12.

53 Asimov, Isaac and Shulman, Jason A., *Isaac Asimov's Book of Science and Nature Questions*, Weidenfeld & Nicolson, New York, 1988, p281.

54 Snively, W.A., "Science and Theology", *The Biblical Illustrator*, Bible Hub, from: http://biblehub.com/sermons/auth/snively/ science_and_theology.htm

55 Keller, Helen, *Let Us Have Faith*, Doubleday, Doran and Co. Inc., 1940.

About The Author

Dr. Paul Ashley has served as a pastor and has taught courses in Bible science, Bible history, and church history as well as Jewish culture, having been a student and teacher of the Text for over 30 years. He has been a frequent speaker on these subjects across the U.S.

He is also a distinguished internationally known scientist with over 35 years of service including Deputy Director of a research laboratory for missile development. He has authored over 200 publications and presentations as well as numerous patents. He is a graduate of Baylor University (BS 1974) and Washington University (MA 1976, MS 1977, and D.Sc. 1978).

www.ingramcontent.com/pod-product-compliance
Lightning Source LLC
Chambersburg PA
CBHW021202020426
42331CB00003B/170